Primer

Instruction Manual

By Steven P. Demme

Math·U·See

1-888-854-MATH (6284)
www.MathUSee.com

Math·U·See

1-888-854-MATH (6284)

www.MathUSee.com

Copyright © 2010 by Steven P. Demme

Graphic Design by Christine Minnich.

Printed in the United States of America

Primer

Math·U·See

SCOPE & SEQUENCE

Math-U-See is a complete and comprehensive K-12 math curriculum. While each book focuses on a specific theme, Math-U-See continuously reviews and integrates topics and concepts presented in previous levels.

Primer

α Alpha | Focus: Single-Digit Addition and Subtraction

β Beta | Focus: Multiple-Digit Addition and Subtraction

γ Gamma | Focus: Multiplication

δ Delta | Focus: Division

ε Epsilon | Focus: Fractions

ζ Zeta | Focus: Decimals and Percents

Pre-Algebra

Algebra 1

Stewardship*

Geometry

Algebra 2

Pre Calculus with Trigonometry

Calculus

*Stewardship is a biblical approach to personal finance. The requisite knowledge for this curriculum is a mastery of the four basic operations, as well as fractions, decimals, and percents. In the Math-U-See sequence these topics are thoroughly covered in Alpha through Zeta. We also recommend Pre-Algebra and Algebra 1 since over half of the lessons require some knowledge of algebra. Stewardship may be studied as a one-year math course or in conjunction with any of the secondary math levels.

Five Minutes for Success

Welcome to *Primer*. I believe you will have a positive experience with the unique Math-U-See approach to teaching math. These first few pages explain the essence of the methodology which has worked for thousands of students and teachers. I hope you will take five minutes and read through these steps carefully.

If you are using the program properly and still need additional help, you may visit Math·U·See online at MathUSee.com/support, or call us at 888-854-6284 (homeschools and individuals) or 800-454-6284 (schools and special education departments).

—**Steve Demme**

Special Instructions for *Primer*

Alpha is the beginning of formal Math-U-See instruction where we emphasize mastery learning. In *Primer* we are giving students a taste of several math concepts, but not requiring them to understand the concepts completely, since they will be presented afresh in successive books. So, make it fun. Use the blocks as much as possible. Let the students develop a positive attitude towards numbers and number concepts as they are introduced to math for the first time. Attitudes are catching. As you and your student build with the blocks, play the games in the lessons, and enjoy the subject, it will set a positive tone for the class. Have fun!

The Goal of Math-U-See

The underlying assumption or premise of Math-U-See is that the reason we study math is to apply math in everyday situations. Our goal is to help produce confident problem solvers who enjoy the study of math. These are students who learn their math facts, rules, and formulas and are able to use this knowledge to solve word problems and real life applications. Therefore, the study of math is much more than simply committing to memory a list of facts. It includes memorization, but it also encompasses learning the underlying concepts of math that are critical to successful problem solving.

THE SUGGESTED 3-STEP MATH-U-SEE APPROACH

In order to train students to be confident problem solvers, here are the three steps that I suggest you use to get the most from Math-U-See *Primer*.

Step 1. Prepare for the Lesson
Step 2. Present the New Topic
Step 3. Practice for the Student

Step 1. Prepare for the Lesson.

Watch the DVD to learn the new concept and see how to demonstrate this concept with the blocks. Study the written explanations and examples in the instruction manual. Many students watch the DVD along with their instructor.

Step 2. Present the New Topic

Present the new concept to your student. Have the student watch the DVD with you, if you think it would be helpful.

a. **Build:** Use the blocks to demonstrate the problems from the worksheet.

b. **Write:** Show the problems on paper as you build them with the blocks.

c. **Say:** Explain the *why* and *what* of math as you build and write.

By using "Build, Write, and Say" (also explained on the DVD) you are helping the students to use their eyes, ears, and hands to learn. Young children think differently than adults. They need to see, touch, and build with concrete objects to order to understand. They take in information through their senses. Dr. Milt Uecker puts it this way: "Children have to move their muscles through every concept."

Do as many problems as you feel are necessary until the student is comfortable with the new material. One of the joys of teaching is hearing a student say *"Now I get it!"* or *"Now I see it!"*

Step 3. Practice for the Student.

Using the examples and the lesson practice problems from the student text, have the students practice the new concept. Coach them through the building, writing, and saying process.

Do as many of the lesson practice pages as necessary for the child to be comfortable, and then proceed to the systematic review pages.

Length of a Lesson

So how long should a lesson take? This will vary from student to student and from topic to topic. You may spend a day on a new topic, or you may spend several days. There are so many factors that influence this process that it is impossible to predict the length of time from one lesson to another. If you move from lesson to lesson too quickly, the student will become overwhelmed and discouraged. If you move too slowly, your student may become bored and lose interest in math. I believe that as you regularly spend time working along with your student, you will sense when to proceed to the next topic.

Confucius is reputed to have said, "Tell me, I forget; Show me, I understand; Let me do it, I will remember."

ONGOING SUPPORT AND ADDITIONAL RESOURCES

Welcome to the Math-U-See Family!

Now that you have invested in your children's education, I would like to tell you about the resources that are available to you. Allow me to introduce you to our staff, our ever improving website, the Math-U-See blog, our new free e-mail newsletter, and other online resources.

Many of our customer service representatives have been with us for over 10 years. What makes them unique is their desire to serve and their expertise. They are able to answer your questions, place your student(s) in the appropriate level, and provide knowledgeable support throughout the school year.

Come to your local curriculum fair where you can meet us face-to-face, see the latest products, attend a workshop, meet other MUS users at the booth, and be refreshed. We are at most curriculum fairs and events. To find the fair nearest you, click on "Events" under "E-sources."

The **Website**, at www.MathUSee.com, is continually being updated and improved.It has many excellent tools to enhance your teaching and provide more practice for your student(s).

ONLINE DRILL

Let your students review their math facts online. Just enter the facts you want to learn and start drilling. This is a great way to commit those facts to memory.

Math-U-See Blog

Interesting insights and up-to-date information appear regularly on the Math-U-See Blog. The blog features updates, rep highlights, fun pictures, and stories from other users. Visit us and get the latest scoop on what is happening.

Email Newsletter

For the latest news and practical teaching tips, sign up online for the free Math-U-See e-mail newsletter. Each month you will receive an e-mail with a teaching tip from Steve as well as the latest news from the website. It's short, beneficial, and fun. Sign up today!

Online Support

You will find a variety of helpful tools on our website, including corrections lists, placement tests, answers to questions, and support options.

For Specific Math Help

When you have watched the DVD and read the instruction manual and still have a question, we are here to help. Call us or click the support link. Our trained staff are available to answer a question or walk you through a specific lesson.

Feedback

Send us an e-mail by clicking the feedback link. We are here to serve you and help you teach math. Ask a question, leave a comment, or tell us how you and your student are doing with Math-U-See.

Our hope and prayer is that you and your students will be equipped to have a successful experience with math!

Blessings,

Steve Demme

Number Recognition
Counting from 0-9

The two skills needed to function in the decimal system are the ability to count to nine, and an understanding of *place value*. In the *decimal system*, where everything is based on 10 (deci), you count to nine and then start over. To illustrate this, count the following numbers slowly: 800, 900, 1,000. We read these as eight hundred, nine hundred, one thousand. Now read these: 80, 90, 100. These are read as eighty, ninety, one hundred. Once you can count, work on place value. The two keys are learning the counting numbers zero through nine, which tell us how many, and understanding place value, which tells us what kind.

When counting, begin with zero, and then proceed to nine. Traditionally we've started with one and counted to 10. Look at the two charts that follow, and see which is more logical.

```
 1  2  3  4  5  6  7  8  9 10      0  1  2  3  4  5  6  7  8  9
11 12 13 14 15 16 17 18 19 20     10 11 12 13 14 15 16 17 18 19
21 22 23 24 25 26 27 28 29 30     20 21 22 23 24 25 26 27 28 29
```

The second chart has all single digits in the first line, and then in the second line, each number in the units place is preceded by a one in the tens place. In the third line, each number in the units place is preceded by two in the tens place. The first chart, though looking more familiar, has the 10, the 20, and the 30 in the wrong lines. When counting, always begin with zero, count to nine, and then start over.

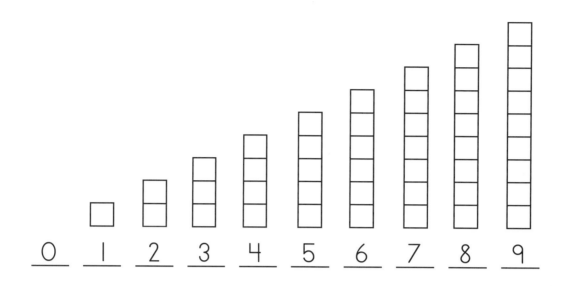

On the worksheets, the squares are the same size as the green unit blocks, so you can place the bars on the paper, count, and then circle the correct number. When there are objects and pictures, count those, and then circle the correct number. You can also have the student trace over the numeral he circled to prepare him for writing. Remember to begin with zero when counting. Notice the senses involved: see the number of squares (visual), hear the correct number as you count (auditory), and build with the blocks (kinesthetic and tactile).

One of the reasons we use blocks is so that students recognize that numbers represent real things. On the worksheets, you will notice that we move from blocks to counting real items like frogs, crayons, shoes, etc.

Example 1

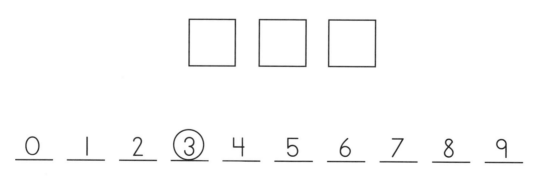

Example 2

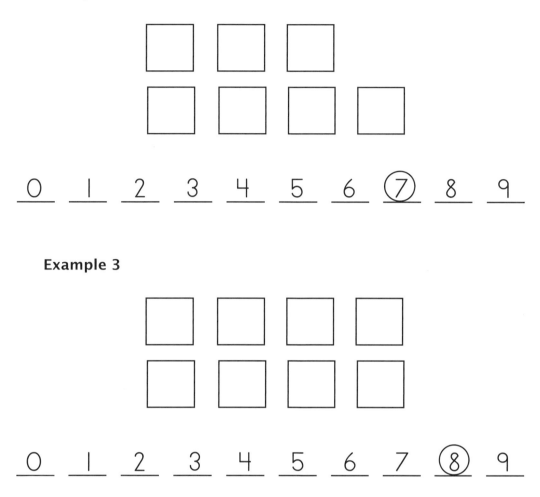

0   1   2   3   4   5   6   ⑦   8   9

Example 3

0   1   2   3   4   5   6   7   ⑧   9

There are solution pages at the back of this instruction manual. The solutions for pages A and D are shown for each lesson.

Writing Numerals

These exercises have two choices. The student is given the number of blocks, and must choose the correct numeral that corresponds to the number of blocks. Take the green unit blocks and place them on the squares in the book. Then circle the correct answer and cross out the incorrect answer. Trace the correct answer and say it as shown in examples 1 and 2.

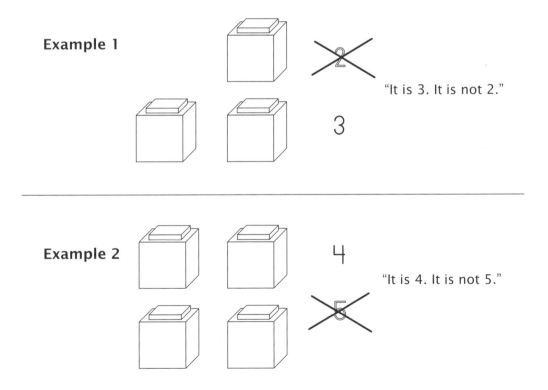

Example 1

"It is 3. It is not 2."

Example 2

"It is 4. It is not 5."

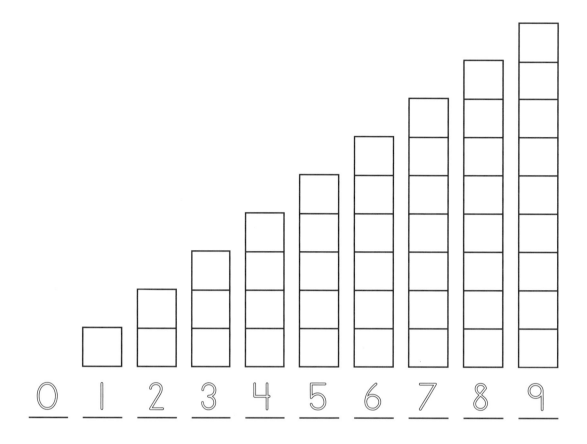

You may choose different ways to have your student write his numerals. The "9" and the "6" may have straight lines down instead of curves connected to the circle. Please teach your student that there is more than one way to write the numeral "4." It may have a closed top, as in most books, or it may be open at the top as we have written it on the worksheets.

In this lesson, practice writing the numerals from 0 to 9. If your student is not ready for this use of his fine motor skills, you may continue by doing the lesson verbally and sharpen counting skills while you write in the correct answers.

Another exercise that may help familiarize your student with his numerals is to use 3 x 5 note cards, or something similar. Write a number on each card, and put the correct number of squares on each one corresponding to the number. Color them if you wish.

Then, when the numbers have been learned, have the student put them in sequential order. Encourage the student to say the numbers aloud as he or she plays with the cards.

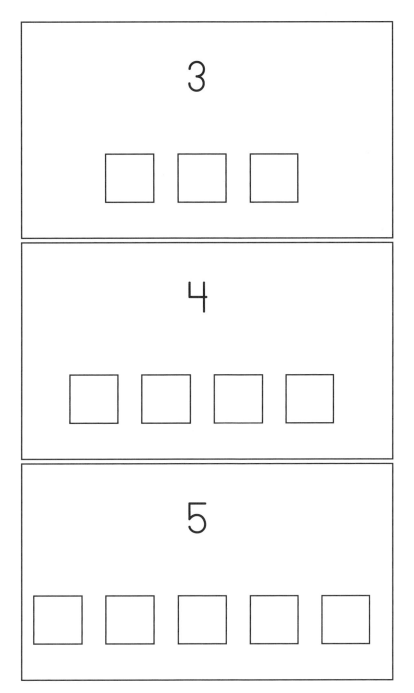

Number Recognition and Writing Numerals

Now we are given the numeral and are asked to color or shade the correct number of squares. The shaded squares don't have to be in any particular spot on the grid. We are focusing on *how many*, not *where*. Study the examples to see how to do these. Place the green unit blocks in the squares before shading or coloring.

Continue to practice writing the numbers by tracing over the outlines of the numerals.

Example 1

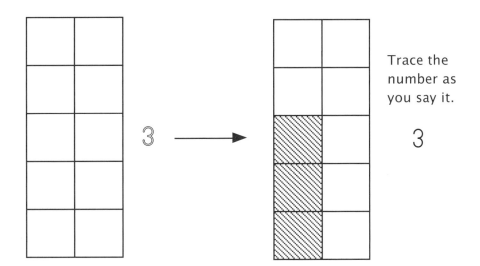

Trace the number as you say it.

3

Example 2

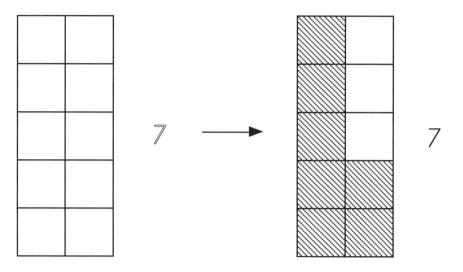

A good place to reinforce number recognition and counting is on a *calendar*. During the course of the day point out the number of the date, and notice the date before and the date following.

1	2	3	4	5	6	7
8	9					

At this level, we are focusing on counting from zero to nine. Later, we will add the larger numbers. Anytime the student is ready, mention that there are seven days in a week. We will study the different numbers of days in the months when we come to ordinal numbers.

1	2	3	4	5	6	7
8	9	10	11	12	13	14
15	16	17	18	19	20	21
22	23	24	25	26	27	28

Geometric Shapes: Rectangles

A *rectangle* is another way of saying right angle. ("Rect" comes from a German word that means right.) At this age, we define a rectangle as a four-sided shape with four square corners. Use the rectangles shown to reinforce counting and familiarize the student with this new shape. Have the student count, write, and say the correct answer as you ask the questions.

1. How many rectangles are there with stripes? (5)
2. How many black rectangles are there? (2)
3. How many white rectangles are there? (5)
4. How many gray rectangles are shown? (4)

Look around to discover examples of rectangles.

Number Recognition and Writing Numerals

These lessons are the opposite of lesson 3. The shaded squares are given, and the student is being asked to write the correct numeral. If the writing skills are not developed enough for this yet, write two possibilities and have the student choose and trace the correct answer. Or you may write the number after he or she tells you what the answer is.

Ask the student to place the correct number of green unit blocks over the shaded squares, count the blocks, and say the number as it is being written.

Example 1

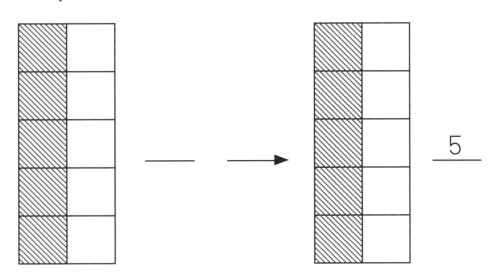

Example 2

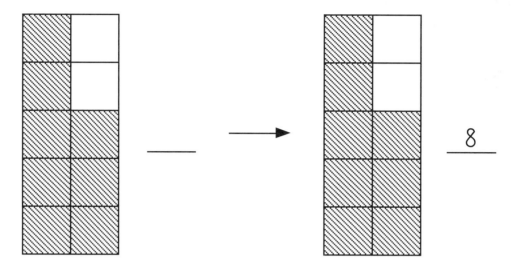

Geometric Shapes: Circles

Introduce *circles*. Look for circles around you, and draw some. Use a coin, the top of a jar, or a plate to construct circles.

1. How many circles have lines in them? (4) ⊕
2. How many circles are black? (5) ●
3. How many white circles are there? (7) ○

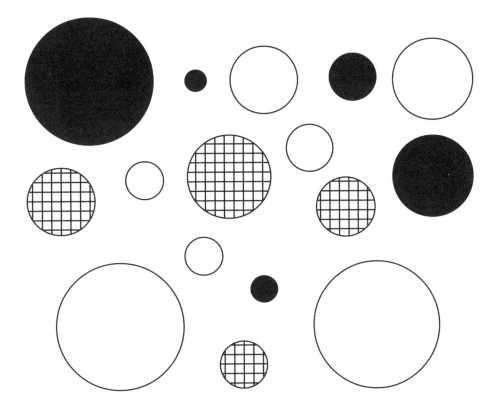

Number Recognition and Writing Numerals

This lesson combines the previous lessons. If the numeral is given, the student reads it and shades the correct number of squares. If the shaded squares are given, the student writes and says the corresponding numeral. Here are some examples.

Example 1

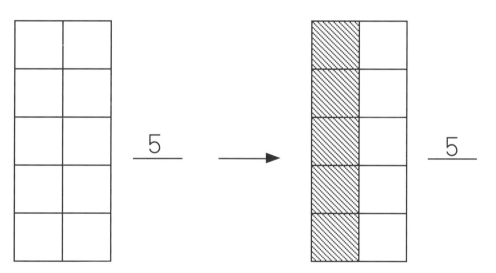

Example 2

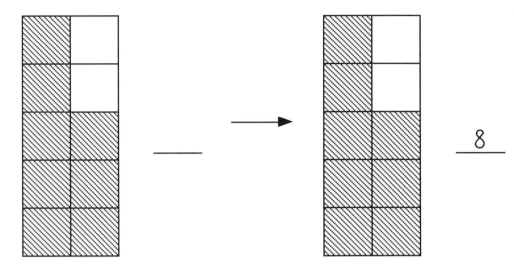

$$\underline{8}$$

Geometric Shapes: Triangles

Introduce *triangles*. "Tri" means three and "angle" means angle: three sides and three angles. Look for triangles in the immediate environment.

1. How many triangles have stripes? (4) ⧄
2. How many triangles are white? (8) △
3. How many gray triangles are there? (7) ◭

LESSON 9

Place Value: Units and Tens

I define this important subject as "Every value has its own place!" To an older child I would add, "Place determines value!" Both are true. There are 10 symbols to tell you how many, and many places to represent what kind or what value. Zero through nine tell us how many; *units* and *tens* tell us what kind. For the sake of accuracy, units will be the word used to denote the first value, instead of ones. One is a counting number that tells us how many, and units is a place value that denotes what kind. This will save potential confusion by not saying "10 ones" or "one ten." Remember, one is a number and units is a place value. The numerals (0, 1, 2, . . . 9) tell us how many tens or how many units. We begin our study focusing on the units and tens, but there are other values, such as hundreds, thousands, millions, billions . . .

When teaching this, I like to use a street since I'm talking about a place. I call the street *Decimal Street* and have the house for the little green units and the house for the tall blue tens next door to each other. We don't want to forget that when counting we only count from zero to nine and then start over. To make this more real, begin by asking, "What is the largest number of units that can live in this house?" You can get any response to this question from zero to nine, and you might say "yes" to all of them, but remind the student that the largest number is nine! So we imagine how many little green beds, or green toothbrushes, or green chairs there would be in the house. Ask the student what else there would be nine of. Do the same with the tens. Remember that in the Units house, all the furniture will be green, and in the Tens house it will be blue like the blue ten bars.

There are directions for making a Decimal Street poster in lesson 10.

Decimal Street

Throughout the program, whenever we teach we will employ the following strategy: Build—Write—Say. To teach place value, we will build the number, count how many in each place, write the number, and then read what we've written.

Let's build 42 (four tens, two units). After building, count how many are "at home" in each house. I like to imagine going up to the door of each home and knocking to see how many are home in each place. Then write the numeral 42 as you count (always beginning with the units) to show the value on paper. Finally, say, "Four tens and two units or forty-two." Build another one and have the student write how many are home. When he or she understands this, you write how many on paper, and have the student build it! Try 37. After he builds it, you read what he has built. Keep practicing back and forth with the teacher building and the student writing and vice versa.

Here is another exercise I do to reinforce the fact that every value has its own place. I have the student close his or her eyes as I move the pieces around by placing the blue tens where the units should be and vice versa . I then ask the student to make sure the blocks are all in the right place. You might call this "scramble the values" or "walk the blocks home." As the student looks at the problem and begins to work on it, I ask, "Is every value in its own place?"

You've probably noticed the important relationship between language and place value. Consider 42, read as "forty-two." We know that it is four blue ten bars (for-ty, "ty" for ten) and two units. When pronouncing 90, 80, 70, 60, and 40, work on enunciating clearly, so that 90 is ninety, not "ninedee." 80 is eighty, not "adee." When you pronounce the number accurately, not only will your spelling improve, but your understanding of place value will improve as well. The number 70 (seventy) is seven tens; 60 (sixty) is six tens. The number 40 is pronounced correctly but spelled without the "u" in it. Carrying through on this logic, 50 should be pronounced "fivety" instead of "fifty." The numbers 30 and 20 are similar to 50; not completely consistent but close enough so we know what they mean. The teens are the real problem.

Some researchers have concluded that one of the chief differences between western (American and Canadian) and eastern (Chinese and Japanese) students is their understanding of place value. The culprit, in the researchers' eyes, is the English language. In eastern culture, when a child can count to nine, with a few minor variations, he can count to 100. This is not so in English with such numbers as 10, 11, 12 and the rest of the teens. Not only are these numbers difficult to teach because there doesn't seem to be rhyme nor reason for their origin, but more importantly, they do not reflect and indicate place value. To remedy this serious deficiency, I'm suggesting a new way to read the numbers 10 through 19. You decide whether this method reinforces the place value concept and restores logic and order to the decimal system.

Ten is "onety," 11 is "onety-one," 12 is "onety-two," 13 is "onety-three," . . . 19 is "onety-nine." Now it is not that the student can't say "ten," "eleven," "twelve," but learning this method enhances his understanding and makes math logical again. Also, students think it is neat.

When presenting place value, or any other topic in this curriculum, model how you think as you solve the problems. As you the teacher work through a problem with the manipulatives, do so verbally, so that as the student observes, he also hears your thinking process. Then record your answer.

Example 1

given visually

As you look at the picture, slowly say it, proceeding from left to right, "fifty-three." Then count, beginning with the units, "1-2-3" and write a "3" in the units place. Then count the tens, "1-2-3-4-5" and write a "5" in the tens place. Do several of these, and then give the student the opportunity to do some.

Example 2
74 (given the written number)

Read the number seventy-four, and then say "seven-ty or seven tens" and pick up seven blue ten bars. Then say "four" and pick up four green unit pieces. Place them in the correct place as you say, "Every value has its own place." Do several of these, and then give the student the opportunity to do some.

Example 3
"sixty-five" (given verbally)

Read the number out loud slowly. Then pick up six blue ten bars as you say, "six-ty" or "six tens." Then say "five" and pick up five green unit pieces. Place them in the correct place as you say, "Every value has its own place." Then write the number 65. Do several of these, and then give the student the opportunity to do some.

Game for Place Value

Pick a Card - Make up a set of cards with 0 through 9 written in green. Then make another stack of cards written in blue with the same numbers 0 through 9. Shuffle the green cards, pick one, and show that number of green unit blocks. If a child picks a green 4, count out four green unit blocks and show them.

When the child is proficient at this game, try it with the blue cards and do the same as before, except choose blue ten bars instead of the green unit blocks. When the student can do the tens well, add the green cards to the blue cards. Have the child choose one card from the green pile and one card from the blue pile and then pick up the correct number of blue ten bars and green unit blocks.

Place Value: Hundreds

This lesson introduces the *hundreds* place. There is one new house on Decimal Street and it is a BIG one. I call it the Hundreds castle. Place value is so important in the base 10 system that we can afford to spend extra time with it. Directions for making Decimal Street are at the end of the lesson, as well as games to reinforce place value. Colored pencils may be easier to use than crayons when doing the worksheets.

If this is too much information or confusing for the student, skip to the next lesson. This and all topics in *Primer* will be taught again in succeeding books.

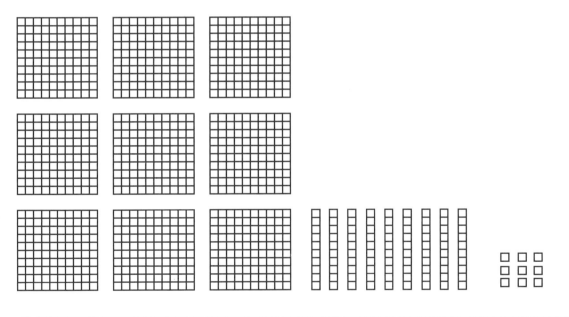

Decimal Street

An important person that I haven't mentioned is Mr. Zero (0). He is a place holder. Let's say you were walking down Decimal Street and knocking on each door to see who was home. If you knock on the Units door and three units answer, then you have "3" in the units place. Next door, at the Tens house you knock and no one answers! Yet you know someone is there feeding the gold fish and taking care of the bird.

This is similar to when a family goes on vacation. Mr. Zero won't answer the door because he's not a ten, but he's the one who holds their place until the tens come home. Upon knocking at the Hundreds big red castle, you find that two hundreds answer the door. Your numeral is thus: 203.

Sometime you will want to mention that even though we begin at the units end of the street and proceed right to left, from the units to the hundreds, when we read the number we do it left to right. We want to get into the habit of counting units first, so that when we add, we will add units first, then tens, and then hundreds.

Example 1

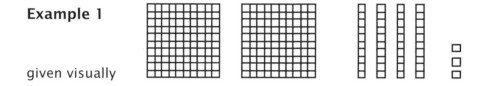

given visually

As you look at the picture, slowly say it, proceeding from left to right, "two hundred forty-three." Then count, beginning with the units, "1-2-3" and write a "3" in the units place. Then count the tens, "1-2-3-4" and write a "4" in the tens place. Lastly, count the hundreds, "1-2" and write a "2" in the hundreds place. Do several of these, and then give the student the opportunity to do some.

Example 2
274 (given the written number)

Read the number "two hundred seventy-four," and then say "two hundreds" as you pick up two red hundred squares. Say "seven-ty or seven tens" and pick up seven blue ten bars.

Then say "four" and pick up four green unit pieces. Place them in the correct place as you say, "Every value has its own place." Do several of these, and then give the student the opportunity to do some.

Example 3
"one hundred sixty-five" (given verbally)

Read the number out loud slowly. Then, as you pick up one red hundred square, say "one hundred." Then say "six-ty" or "six tens" and pick up six blue ten bars. Then say "five" and pick up five green unit pieces. Place them in the correct place as you say, "Every value has its own place." Then write the number 165. Do several of these, and then give the student the opportunity to do some.

Games for Place Value

Pick a Card - This game was described in lesson 9. For this lesson, create one more stack of cards with the numbers 0 through 9 written on them in red.

When the child is proficient playing with the green and blue cards, add in the red cards and proceed as before. Place them in three stacks, shuffle, and draw from each stack. Have the student use the blocks to show you the number he has drawn. Example: You select a green 4, blue 7, and red 2. Choose 2 red hundreds, 7 blue tens, and 4 green units. You may want to limit the number of red number cards to correspond to how many hundreds you have.

Feel free to make variations. If you have a student who has difficulty writing numbers, use the cards until the fine motor skills are more developed.

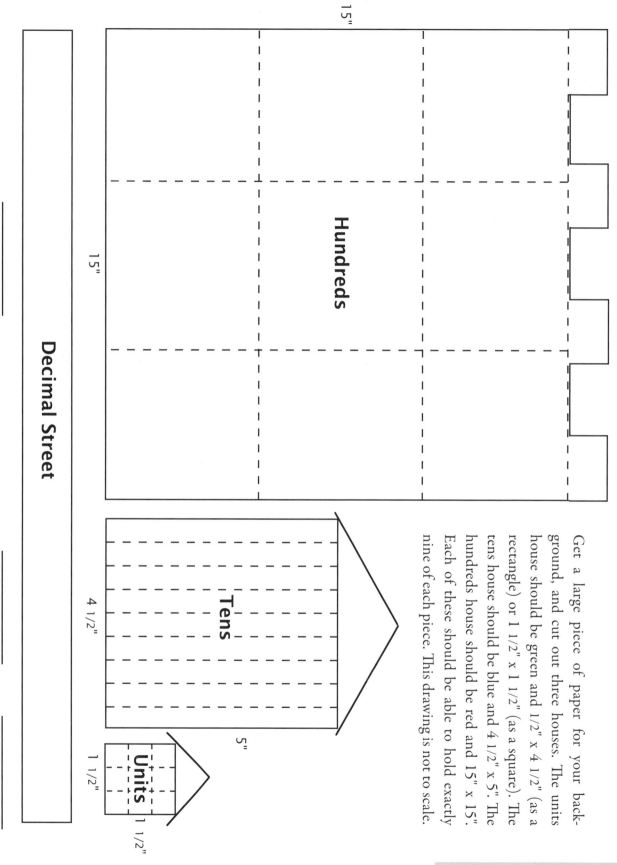

Decimal Street

Hundreds

15"

15"

Tens

4 1/2"

5"

Units

1 1/2"

1 1/2"

Get a large piece of paper for your background, and cut out three houses. The units house should be green and 1/2" x 4 1/2" (as a rectangle) or 1 1/2" x 1 1/2" (as a square). The tens house should be blue and 4 1/2" x 5". The hundreds house should be red and 15" x 15". Each of these should be able to hold exactly nine of each piece. This drawing is not to scale.

Unit Bars

Before attempting addition, we want to teach the student that a group of three green units is the same as one pink unit bar, which is three units long. Place the blocks side by side to teach this. Show that if the three individual units are "glued" together, they are the same length as one pink unit bar. The skill we are teaching here is often called the conservation of matter. A young child would rather have five pennies than one nickel. His brain has developed to the next level when he knows that five pennies and one nickel have the same value.

This is very important, as this is what addition is: two units plus three units being the same as five units.

So in terms of the blocks, wean the student from using the green units exclusively to using the colored unit bars as well. On the worksheets, match the green ones on the left with the correct unit bar on the right.

You may also have him or her color the bars on the right to correspond to the green unit pieces on the left.

There are four levels of combining or arithmetic.

1.   Counting.

2.   Adding, which is fast counting.

3.   Multiplying, which is fast adding.

4.   Exponents, which are fast multiplying.

To move from counting to adding, the student has to learn the unit bars, so that we add two plus three and get five, instead of counting "1–2," and then "3–4–5."

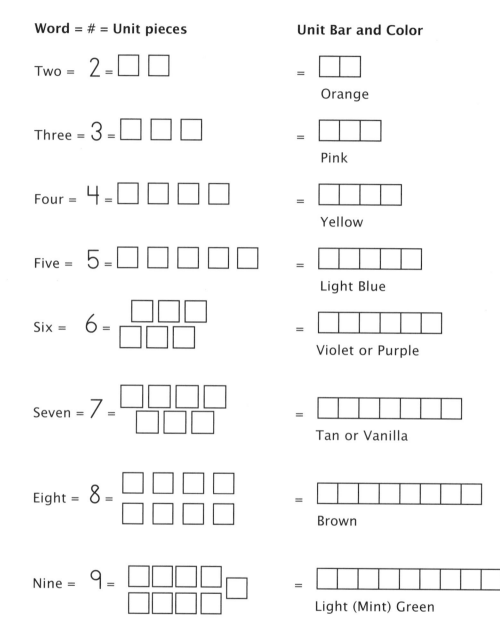

Word = # = Unit pieces

Two = 2 = ☐ ☐

Three = 3 = ☐ ☐ ☐

Four = 4 = ☐ ☐ ☐ ☐

Five = 5 = ☐ ☐ ☐ ☐ ☐

Six = 6 = ☐☐☐ / ☐☐☐

Seven = 7 = ☐☐☐☐ / ☐☐☐

Eight = 8 = ☐☐☐☐ / ☐☐☐☐

Nine = 9 = ☐☐☐☐☐ / ☐☐☐☐

Unit Bar and Color

= ☐☐
Orange

= ☐☐☐
Pink

= ☐☐☐☐
Yellow

= ☐☐☐☐☐
Light Blue

= ☐☐☐☐☐☐
Violet or Purple

= ☐☐☐☐☐☐☐
Tan or Vanilla

= ☐☐☐☐☐☐☐☐
Brown

= ☐☐☐☐☐☐☐☐☐
Light (Mint) Green

Games for Unit Bar Identification

Simon Says - Examples: "Put a three on your nose" or "Hide two fives in a pocket."

What's Missing? - Start with the one through nine blocks. Ask the student to hide his eyes while you remove one of the blocks. Take turns. The options to this are numerous. Try removing two blocks at a time or starting with two of each number and removing one or two blocks.

The Grab Bag - Put the one through nine blocks in an opaque bag. Take turns either drawing a number card and finding a particular block or telling each other to feel around and find a certain block. The easier version is to simply name the one you are about to pull out. A harder version would be to name the missing block after one has been removed.

Blocks and Symbols Match Up - Make a set of 3 x 5 cards with the symbols 1 through 9 on them. On the back of the cards, draw a colored dot to match the blocks. Place the cards symbol side showing, and match the blocks to them. This is especially good as a shelf activity that students can get out on their own. It is self-correcting. Go ahead and let them peek; they stop soon enough. Introduce the symbols by doing the first three, then the next three, and finally all nine.

Sing and Grab - "If you're happy and you know it . . ." Clap three times, or grab a five, etc.

On the worksheets, match the correct answer on the right with the problem on the left.

Example 1
Use the blocks. Count, match, and color. Write the number and say it.

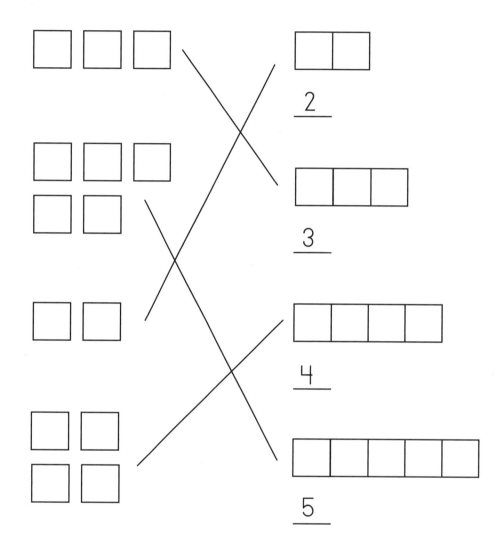

Addition: Introduction and Symbolism

As we begin *addition*, think of it as smooshing two blocks together and then finding out which block is the same length as the result. If we smoosh a five bar and a three bar, it is the same length as an eight bar. We build it first, and then write the symbols 5 + 3 = 8. When learning the *plus sign*, which is the symbol for addition, maybe a train would help to illustrate the concept. When we take our "5 car" and hook it with the "3 car," the plus sign is the symbol for the connector and the result is an "8 car." To make it real to a student, perhaps you could say, "If you have five dogs now, and the neighbor gives you three more dogs, how many dogs do you have?" The answer is eight.

Even though the picture is sufficient in many cases, the more you use the blocks, the better will be the understanding on the part of the student.

In the student book, color the unit bars to make the connection from the blocks to the picture. Then write the correct numbers on the lines below the picture, and read the equation out loud.

Example 1
Solve 5 + 3 =

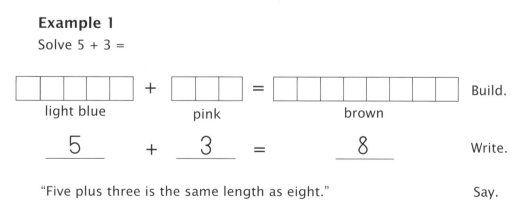

light blue	pink	brown	Build.
5	+ 3	= 8	Write.

"Five plus three is the same length as eight."   Say.

Place the five bar and the three bar on top of, or beside, the eight bar. Notice that with the bars arranged this way, the two lines make an *equals sign*.

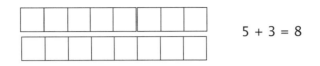

5 + 3 = 8

Example 2

Solve 2 + 7 =

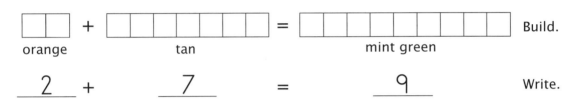

orange   tan   mint green   Build.

___2___ + ___7___   =   ___9___   Write.

"Two plus seven is the same length as nine."   Say.

2 + 7 = 9

Addition: +1

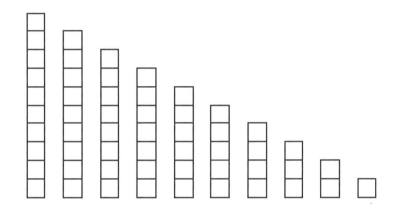

You can use the following as a narrative and adapt it until the concept is learned. Arrange the blocks as shown above. Have the student put his/her finger on the three bar. Say "Point to the number that is *one more* than this." (four) "What is *one larger* than that?" (five) "If you *add one* to five what do you have?" (six) "What is *one greater* than six?" (seven) "If I add one to seven what do I have?" (eight) "What is one more than eight?" (nine) Practice until the student knows one more than. Change your vocabulary to teach these important words that indicate addition.

When showing the one facts, begin by taking the green unit bar and placing it end to end with the pink three bar. Ask the student if he can find another unit bar that is the same length as the one bar and the three bar smooshed together. This is addition: placing a bar end to end with another bar and finding a third bar that is the same length, in this case the yellow four bar. Beside the bars write: 1 + 3 = 4 while saying, "One plus three is the same length as four" or "One plus three equals four." The child sees it, builds it, writes it, reads it, and hears it. After this barrage, hopefully he understands it and is on his way to remembering it.

Example 1

Solve 3 + 1 =

Place the 3 bar and the 1 bar
on top of, or beside, the 4 bar.

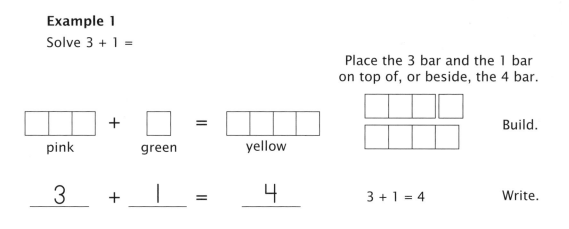

Build.

___3___ + ___1___ = ___4___   3 + 1 = 4   Write.

"Three plus one is the same length as, or equal to, four."   Say.

Counting to 20

The transition from 9 to 10 is pivotal in understanding counting to 100 and regrouping (carrying). Work with your student in counting to 20. On the next page there is a story to make counting to 20 more fun and hopefully more understandable.

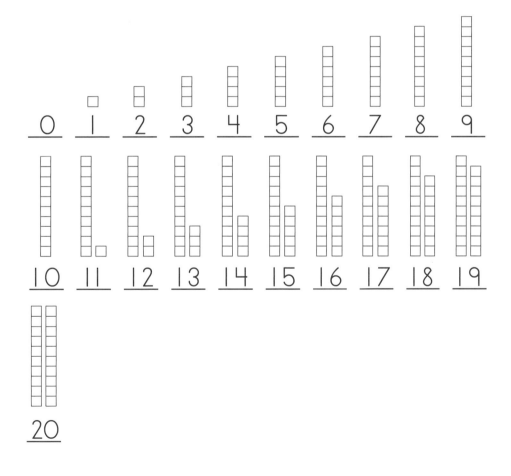

On the worksheets, the students practice writing their numbers on the lines that correspond to the pictures of the blocks. I left a few numbers in just to make it interesting.

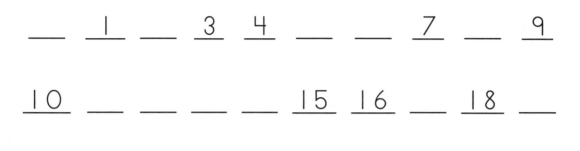

___ 1 ___ 3 4 ___ ___ 7 ___ 9

10 ___ ___ ___ ___ 15 16 ___ 18 ___

20

Another way to teach counting to 20 uses the first two houses in Decimal Street. Begin by placing one green unit block in the units house, and then say and write "1." Add another unit block, and say and write "2." Keep doing this procedure until there are nine blocks at home in the units house. Now try to add one more, and notice that there is no more room. The dry math reality at this point is that the 10 individual green units are transformed into a blue ten bar. To make it more interesting, we can make up a story of these new units who keep moving into our neighborhood.

Apparently they heard that we have a nice home and naturally want to live with us. After we welcome them into our home, we find that we like them as well. So when the house is full (nine units) and this new fellow wants to live with us, we have a family discussion. We decide that in order for him to live with us, we will all have to become a blue ten bar and move into the house next door, so we can be together. As more units keep arriving, they can stay in the units house because it is now open. When the first unit arrives after we have moved into the tens house, we have one ten bar and one unit bar, or 11. The next unit arrives, and there are 12, or one ten and two units, living on our street. Eventually we have 19 units, and when the next unit arrives, another ten is formed and moves in with the original ten, so now there are two tens or 20. This is illustrated on the next page.

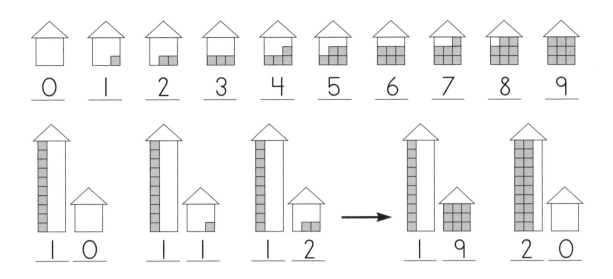

When teaching counting to 20, encourage the student to count aloud using the "nicknames"—one, two, . . . , eight, nine, ten, eleven, twelve, . . . , twenty—as well as the "proper names"—eight, nine, onety, onety-one, onety-two, . . . , two-ty. The nicknames are used most often, but the proper names will help with comprehending place value.

The calendar mentioned in lesson 3 is a good place to apply counting to 20.

Addition: 2 + 2 and 3 + 3; Vertical Addition

There are two new addition facts: 2 + 2 and 3 + 3. Present these with the blocks: building, writing, and saying. When they have been learned, a neat way to reinforce and apply this new skill is by getting dominoes with two and two, as well as three and three, and then asking how many dots are on each face.

You can do the same with two dice with double twos and double threes.

 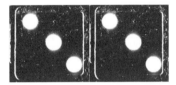

Example 1

Solve 3 + 3 =

Place the 3 bar and the 3 bar
on top of, or beside, the 6 bar.

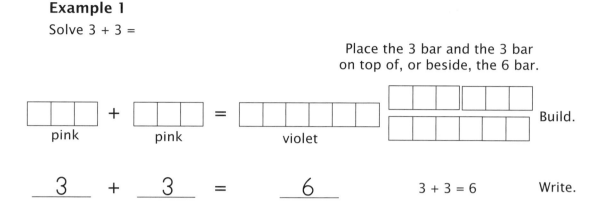

Build.

$$\underline{\quad 3 \quad} + \underline{\quad 3 \quad} = \underline{\quad 6 \quad}$$

3 + 3 = 6   Write.

"Three plus three is the same length as, or equal to, six."   Say.

Teach the student that you may write an addition problem horizontally or vertically. See examples 2 and 3. It may be helpful to show the line separating the problem from the answer with two lines and to read it as "equals." Then explain that you really only need one line when doing it this way.

Example 2

Solve: 1
         + 5

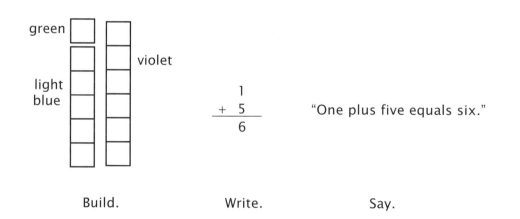

$$\begin{array}{r} 1 \\ + 5 \\ \hline 6 \end{array}$$

"One plus five equals six."

Build.   Write.   Say.

Example 3

Solve:
$$\begin{array}{r} 6 \\ + \ 1 \\ \hline \end{array}$$

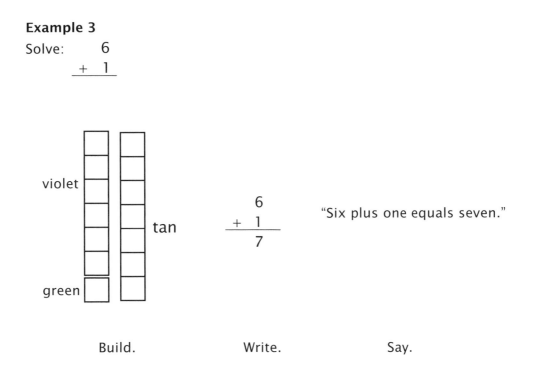

violet

green

tan

$$\begin{array}{r} 6 \\ + \ 1 \\ \hline 7 \end{array}$$

"Six plus one equals seven."

Build.   Write.   Say.

Teaching Tip

I had a friend who taught at a small private school. Each day, she asked her students to identify the date. Then she would ask them how many ways they could express that number. For example, if the date was 7, it could also be 6 + 1, 5 + 2, or 7 + 0. Using the blocks, the student might also discover that 7 could be 3 + 3 + 1. Encourage this kind of exploration.

Shapes: Squares; Addition: 4 + 4 and 5 + 5

A *square* is a special kind of rectangle. It has four right angles, so it is a rectangle, but all four sides of a square are the same length. So a rectangle with all sides the same length is a square. Another way of thinking of it is that a square has four right angles and four sides the same length.

We are going to work on two more addition problems: 4 + 4 and 5 + 5. See examples 1 and 2 on the next page.

Example 1

Solve 4 + 4 =

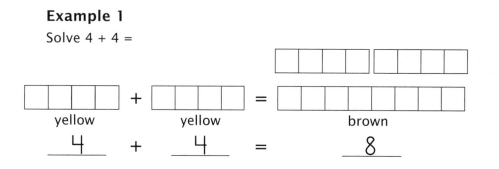

yellow   yellow   brown

    4   +   4   =   8

4 + 4 = 8

Example 2

Solve 5 + 5 =

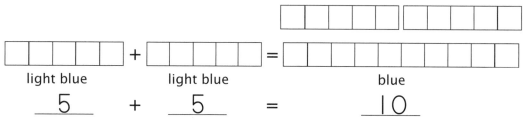

light blue   light blue   blue

    5   +   5   =   10

5 + 5 = 10

Skip Counting by Two

Skip counting is the ability to count groups of the same number quickly. For example, if you were to skip count by threes you would skip the one and the two and say "three," skip the four and the five and say "six," and then say "nine, twelve, fifteen, eighteen," etc. If you were to skip count by sevens, you would count "7–14–21–28–35–42–49–56–63–70."

There are several reasons for teaching skip counting.

1. It lays a solid foundation for learning the multiplication facts. We can write 3 + 3 + 3 + 3 as 3 x 4. If a child can skip count, he could say "3–6–9–12." Then he could read 3 x 4 as "three counted four times is 12." As you learn to skip-count, you are learning all your multiplication facts in order.

2. Skip counting teaches the concept of multiplication. I had a teacher tell me that her students had successfully memorized their facts but didn't know what they had acquired. After she had taught them skip-counting, they understood what they had learned. Multiplication is fast adding of the same number.

3. As a skill in itself, multiple counting is helpful. Skip counting is counting in multiples of a certain number. A pharmacist attending a workshop told me he skip counted when counting pills as they went into the bottles.

4.   It teaches you the multiples of a number, which are so important when making equivalent fractions and finding common denominators. In equivalent fractions 2/5 = 4/10 = 6/15 = 8/20. The numbers 2–4–6–8 are the multiples of two, and 5–10–15–20 are the multiples of five.

In this lesson, we will be counting by twos. When first introducing this, try pointing to each square and, as you count the squares, say the first number quietly, and then ask the student to say the second number more loudly. Continue this practice, saying the first number more quietly each time until you are silently pointing to the first square and pointing to the second square as the student says the number loudly. On the homework sheets, only write numbers in the squares with the lines in them. See examples 1 and 2. In example 3, we are counting geometric shapes in groups of two.

Some other practical examples of counting by twos are eyes, ears, hands, feet, shoes, and socks. You might ask the student to count all the eyes or shoes in the family or in the classroom. This is also a good time to learn the prefix "bi" that denotes "two" in words such as bicycle or biped.

Another way to teach this skill is with the *Skip Count and Addition Song Book*. Included is a CD with the skip-count songs from the twos to the nines sung to popular tunes taken from hymns and Christmas carols.

Example 1
How many boxes are there?

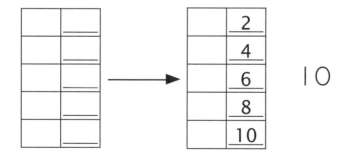

Example 2

How many boxes are there?

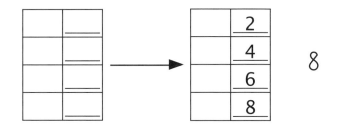

8

Example 3

How many triangles are there?

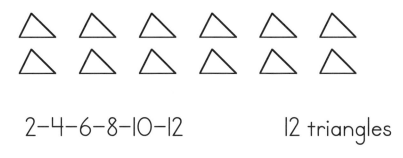

2-4-6-8-10-12   12 triangles

Addition of Tens

So far all of our adding has been units to units. Two units plus two units is four units. The key phrase to remember here is that "to compare or combine, you must be the same kind." Combining is adding. You can only come up with a new number when you add the same kind. For example, three units plus three units is six units. But three units plus two tens is two tens and three units or 23.

We have been adding units plus units so far, but now we are going to add tens plus tens.

Example 1

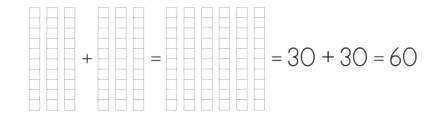

Verbalized: "Three tens plus three tens equals six tens," or "Thirty plus thirty equals sixty."

Example 2

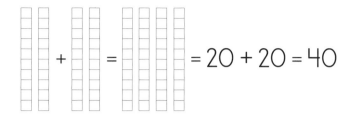 $= 20 + 20 = 40$

Verbalized: "Two tens plus two tens equals four tens," or "Twenty plus twenty equals forty."

Example 3

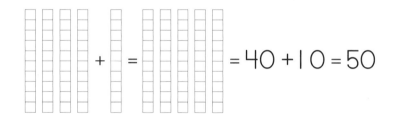 $= 40 + 10 = 50$

Verbalized: "Four tens plus one ten equals five tens," or "Forty plus ten (onety) equals fifty."

Skip Counting by 10

After learning to skip count by two, we need to learn skip counting by 10. Some practical examples are fingers on both hands, toes on both feet, and pennies in a dime. Notice that the multiples of ten—twenty, thirty, forty, etc., all end in "ty." The suffix "ty" represents ten. So six-ty means six tens, and seventy means seven tens.

Example 1
As with the twos, skip count and write the number on the line. Say it out loud as you count and write. Then write the numbers in the spaces provided beneath the figure.

									10
									20
									30
									40
									50
									60

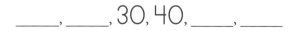

_____, _____, 30, 40, _____, _____

Example 2

Fill in the missing information on the lines.

____, ____, ____, 40, ____, ____, 70, 80, 90, ____

Solution

10, 20, 30, 40, 50, 60, 70, 80, 90, 100

Addition of Hundreds

Recall that "to compare or combine, you must be the same kind." In this lesson, we apply this principle to adding hundreds to hundreds. When doing this vertically as in example 2, make sure your place values are lined up directly beneath one another.

Example 1

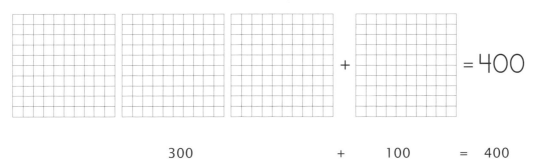

$$300 \qquad + \qquad 100 \qquad = \qquad 400$$

Verbalized: "Three hundreds plus one hundred equals four hundreds," or "300 + 100 = 400."

Example 2

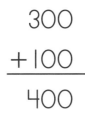

$$
\begin{array}{r}
300 \\
+100 \\
\hline
400
\end{array}
$$

Example 3

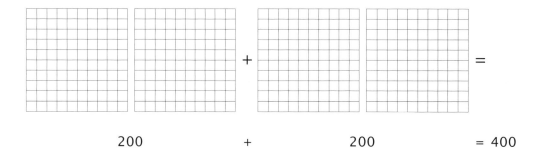

200   +   200   = 400

$$200$$
$$+\ 200$$
$$\overline{400}$$

LESSON 21

Solving for an Unknown

This lesson is a good place to practice counting from 0 to 100.

While teaching addition facts, I like to solve for an unknown for three very important reasons.

1. It reinforces the basic facts.
2. It provides a foundation for subtraction.
3. It familiarizes the students with algebra.

Let's do some examples. Notice that we are not teaching algebra abstractly (just letters and numbers on paper), but concretely with the manipulatives, to give meaning to the letters and numbers. **When solving for the unknown, allow the student to use the blocks for as long as they are needed.**

Example 1

$$\underline{} + 3 = 9$$

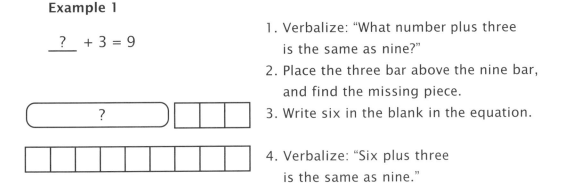

1. Verbalize: "What number plus three is the same as nine?"
2. Place the three bar above the nine bar, and find the missing piece.
3. Write six in the blank in the equation.

4. Verbalize: "Six plus three is the same as nine."

The way you verbalize this is essential to understanding the concept. When you are looking for the unit bar, don't be afraid to experiment. I usually reach for the four bar or the five bar first and try it. My ego can handle being wrong, and I want to show that it is okay to "eliminate possibilities." I like to encourage a student's attempts to experiment until he or she finds the correct solution. Later, we can move to X + 3 = 9. In that case, we are using a letter because we don't know exactly what number makes it the same, or equal, or an equation. I define *algebra* as, "When we don't know what number to use, pick a letter!"

(Some *Primer* students may not be ready for this part. Use your judgement.) Although we are happy that the student can solve for an unknown, we are not satisfied until they can make a word problem out of this equation. After all, the end product of our math instruction is to apply it to real-life situations. For our problem, you might say that you need nine dollars by the weekend and you have three dollars. How many more do you need? Or Charlie Brown has three players. How many more does he need for his baseball team (nine)? While doing the worksheets, encourage the student to make "word problems" or "story problems" to fit the equations. This will help us later when we move in the opposite direction and are are asked to write an equation for a story problem.

Example 2

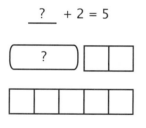

$$\underline{\quad ? \quad} + 2 = 5$$

1. Verbalize: "What number plus two is the same as five?"
2. Place the two bar above the five bar, and find the missing piece.
3. Write three in the blank in the equation.

4. Verbalize: "Three plus two is the same as five."

Games for Problem Solving

Who are you? Who am I? - Example: "Together we are seven, you are a one, so what am I?" When the student has mastered numbers added up to 10, extend the numbers past 10 if he or she is ready.

Both Sides the Same - Get a large piece of white paper and draw a line down the middle. Place a number bar on one side and a smaller bar on the other side, with a missing piece hidden under a bowl or napkin. For example, you place a seven bar on one side, and showing on the opposite side is a three bar and a bowl. Ask what is hidden under the bowl and say, "What plus three is the same as seven?" When the student figures it out, pick up the bowl and reveal the four bar. Start with smaller numbers and progressively place larger amounts on the known side.

Skip Counting by Five

In this lesson, we're teaching skip counting by fives. Use the same techniques to introduce and teach this important skill that have worked before. Some practical examples are fingers on one hand, toes on a foot, pennies in a nickel, players on a basketball team, and sides of a pentagon. We can also remind the student that one nickel has the same value as five pennies, and apply the skill of counting by fives to find out how many pennies are in several nickels.

Example 1

As with the twos, skip count and write the number on the line. Say it out loud as you count and write. Then write the numbers in the spaces provided beneath the figure.

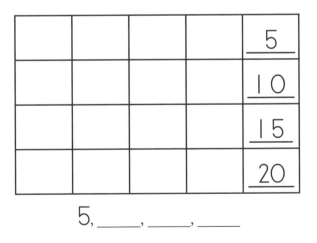

5, _____, _____, _____

Notice that skip counting by five, or fast counting by five, is a way of illustrating multiplication.

The rectangle on the previous page shows five counted four times or 5 x 4, which is 20. We will use this relationship in later books to help the student see the connection between skip counting and multiplication.

Example 2
Fill in the missing information on the lines.

5, ____, 15, ____, ____, 30, ____, 40, 45, ____

Solution

5, 10, 15, 20, 25, 30, 35, 40, 45, 50

Tally Marks

When you use tally marks, you can only use four lines in a row, with one line for "one," two lines for "two," three lines for "three," and four lines for "four." To show "five," you put a slash diagonally through four lines. Tally marks are very useful when keeping track of slowly changing information. Since cowboys used a tally book to keep track of their cattle, I think of cows slowly moving through a gate and a cowboy sitting on a fence, making one mark for each cow passing beneath him. Or perhaps you are on a trip, and you want to count red cars that you pass. Each time you pass a red car, you make a line until you get to the fifth one, and then you make a slash.

Look at the chart below to see how to represent the numbers from 1 to 10. Then study the examples as we change from tally marks to a number and then from a number to tally marks.

1	│	6	∭ │
2	‖	7	∭ ‖
3	‖│	8	∭ ‖│
4	‖‖	9	∭ ‖‖
5	∭	10	∭ ∭

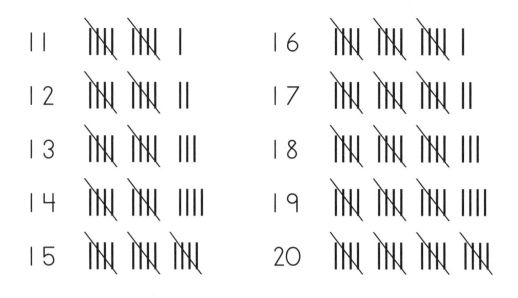

11   𝍷𝍷 𝍷 𝍷   16   𝍷𝍷 𝍷𝍷 𝍷𝍷 𝍷

12   𝍷𝍷 𝍷𝍷 𝍷𝍷   17   𝍷𝍷 𝍷𝍷 𝍷𝍷 𝍷𝍷

13   𝍷𝍷 𝍷𝍷 𝍷𝍷𝍷   18   𝍷𝍷 𝍷𝍷 𝍷𝍷 𝍷𝍷𝍷

14   𝍷𝍷 𝍷𝍷 𝍷𝍷𝍷𝍷   19   𝍷𝍷 𝍷𝍷 𝍷𝍷 𝍷𝍷𝍷𝍷

15   𝍷𝍷 𝍷𝍷 𝍷𝍷   20   𝍷𝍷 𝍷𝍷 𝍷𝍷 𝍷𝍷

Example 1

Change the number 7 to tally marks.

$$7 = 5 + 2 = \;\text{卌}\;\;\text{||}$$

Example 2

Change 卌 卌 |||| to a number.

卌 卌 |||| = 5 + 5 + 4 = 14

Addition: Making 10

There are two ways to teach addition facts. I call the first the logical way, which is used to teach most of the facts The second is the family method, which we use for the 10 family. Take a ten bar and place it on the table or floor. Using two of the colored unit bars, how many ways can you make 10? Be sure you verbally say "two unit bars" because addition facts are the combination of two numbers, not three or four numbers. In figure 1, you can see five ways of making 10. They are 1 + 9, 2 + 8, 3 + 7, 4 + 6, and 5 + 5. Our "10 family" is made up of five different facts. After you build these, write them down and say them to make it multi-sensory.

Figure 1

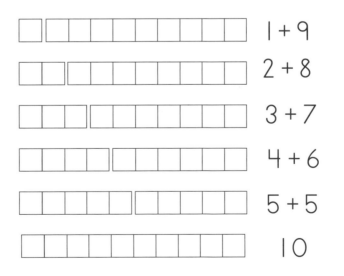

It is important to teach the facts that make 10 using this "10 family" method.

Games for Making 10

Build a Wall - See who can build the highest wall that is 10 units long using two blocks in each row snapped together. After you build your wall, write down the equation that corresponds to the two pieces in each row of the wall. If the first floor is a six bar and a four bar, then the equation is 6 + 4 = 10.

Fill in the Space - Fill in the blank space with the correct block. You are still building on the ten bar, but instead of placing two blocks to be the same length as the ten bar, you place only one and the student places the other one. Then it is his turn to place one, and you figure out the length of the missing piece. Example: Begin with a ten bar, and then choose a seven bar and snap it on top of the ten bar. Have the student find the piece that makes ten, in this case the three bar. Have the student write down 7 + 3 = 10. Then let the student be the teacher and chose a different piece, and you find the missing unit bar and write it down.

Race to a Hundred (and Race You Back) - Use playing cards (face cards removed and aces are ones), the cards you made for Pick a Card, or 10-sided dice. Draw a card, say an eight, and pick up the eight bar and stick it on your hundred square. If your next card is a one or a two, no problem; just stick a unit block or a two bar by the eight bar. If the card is a seven, put the seven bar somewhere else on the hundred square.

On your next turn, pick another number and be on the lookout for blocks that can even out a line, like a two bar to go with the eight bar. Start at zero and accumulate blocks until you reach 100 and cover the entire hundred square.

Race You Back means you don't have to motivate students to put their blocks away! In this game, you go in reverse and take off the value of the card. When they get good at this, require each row of 10 to be filled in from top to bottom as they go. This forces dynamic addition/subtraction to be accomplished. By dynamic, I mean that blocks will have to be "traded in" to get the correct blocks that would fill in the required amount.

Here's an example: the first roll is seven and second roll is a six. The six would be broken into two threes; one three would fill in the first row along with the seven, and the second three would start the next row.

For the race back to zero, I let them pull the rolled number from anywhere within the hundred. A little dynamic work will be needed at the 100 point and when back near zero—go as slowly as necessary. (For example: There is only a five to go to be finished and the student rolls an eight. The eight must be "broken" into a five and a three.)

Example 1

Solve 3 + 7 =

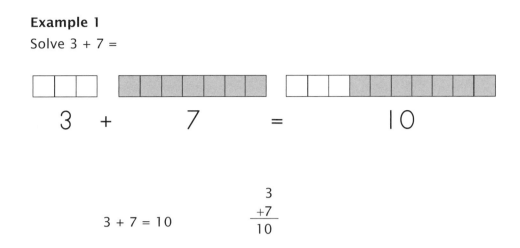

$$3 + 7 = 10$$

$$\begin{array}{r} 3 \\ +7 \\ \hline 10 \end{array}$$

Example 2

Solve 6 + 4 =

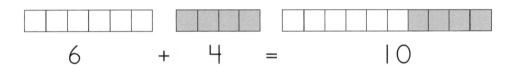

$$6 + 4 = 10$$

$$\begin{array}{r} 6 \\ +4 \\ \hline 10 \end{array}$$

Example 3

Solve 8 + 2 =

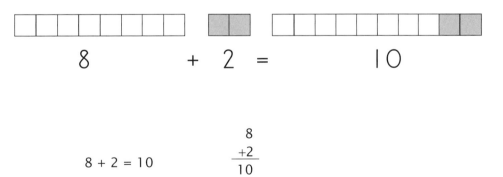

$$8 + 2 = 10$$

$$\begin{array}{r} 8 \\ +2 \\ \hline 10 \end{array}$$

Skip Count to Find Area

In this lesson, the student uses skip counting to find the area of a rectangle. Write the numbers in the spaces where there are lines, as before. Your answer is the last number or largest number you write.

If you get good at this, you can skip count verbally, and then write your answer in the lower right-hand box.

Example 1

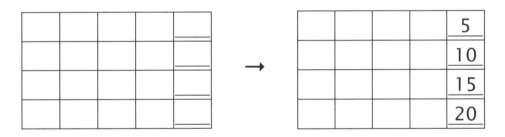

Example 2

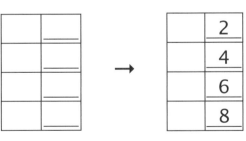

Example 3

									10
									20
									30
									40
									50
									60

Telling Time with Minutes

When the student has mastered skip counting by fives, he is ready to learn how to tell time. This can be challenging with a clock that is not digital. We'll begin by taking six of the ten bars and explaining that there are 60 *minutes* in one hour. Next replace each ten bar with two five bars. If you don't have 12 five bars, use five units, or a four bar and one unit, or a three bar and a two bar. Arrange your 12 groups of five in a circle (which really is a dodecagon, or 12-sided polygon), so you have your 60 minutes in the shape of a clock. Beginning at the top, start skip counting by fives and go around the clock: 5–10–15–20–25–30–35–40–45–50–55–60.

Choose any long unit bar, turn it on its side so that it is smooth, and you have your minute hand. Point it at different areas, and beginning at the top, count your minutes. There is a template to make your clock in the back of the student book. Using all of the blocks in the basic box of blocks, you can make a clock using various combinations for fives (five unit pieces, a two and a three, or a four and a unit).

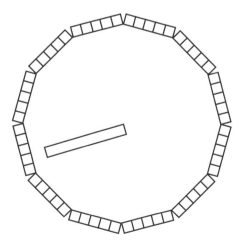

Even though we emphasize skip counting by fives for reading the clock and finding the minutes, you may also remind your student that there are 60 possible minutes, and they are not all multiples of five. When giving problems on your own block clock, move the minute hand to numbers signifying 1, 2, 8, 34 or 51. Reinforce this by having the student count around the clock by ones, saying the multiples of five loudly, as you did when you learned skip counting by fives: 1–2–3–4–5.

To help the student see the progression of the minutes, build several partial clocks as in examples 1 and 2. Count the minutes by beginning at the top and moving around to the right or clockwise.

Example 1
Count the minutes.
5–10, so 10 minutes.

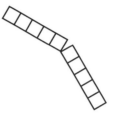

Example 2
Count the minutes.
5–10–15, so 15 minutes.

Telling Time with Hours

When the minutes are mastered, we can add the *hours* by placing a green unit bar at the end of the first five bar (outside the circle) and pointing away from the center of the clock. To distinguish between the minutes and hours, I leave the minutes (unit bars) right side up, but I place the hours upside down with the hollow side showing. You can still see the color and how many hours there are, but it helps to distinguish the minutes and the hours. Place your orange bar (upside down so the hollow side is showing) at the end of the second five bar. Continue this process with all the unit bars through 12. Choose a unit bar smaller than your minute hand for an hour hand. Turn this upside down, so the student makes the connection between the hour hand and the hours, since both are upside down with the hollow side showing.

Now position the hour hand so that it points between the two and the three. This is the critical point for telling time. Is it two o'clock or three o'clock? I've explained this to many children with success by aiming the hour hand at the two and saying, "He just had his second birthday and is now two." Then I move the hand towards the three a little, and ask, "How old is he now? Is he still two?" "Yes." Then I move the hand a little further and ask if he's still two. "Yes." I do this until the hand is almost pointing to three and ask the question, "What about the day before his next birthday; how old is he?" "Still two." He is almost three, but still two. Practice this skill until the student can confidently identify which hour it is by moving the hour hand around the clock.

On the DVD, lessons 27 and 28 are taught concurrently. After you watch the DVD for this lesson, you will know how to teach hours first, and then in lesson 28, teach the hours and minutes together.

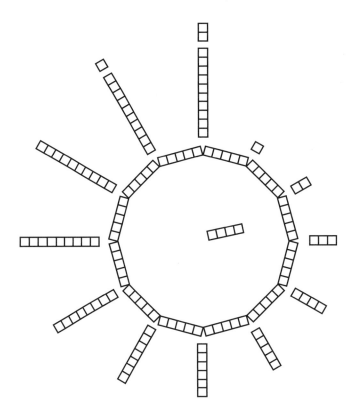

Telling Time with Minutes and Hours

When the hour hand is mastered, put the hours and minutes together. With the hour hand pointing between the three and the four in the example, the hour is three. Skip count by fives beginning at the top to get 30 minutes. Say it as "three thirty" and write it as 3:30.

When this is mastered, have the student tell time using a real clock.

Example 1

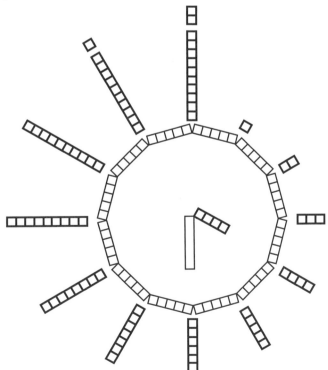

Subtraction: Introduction and Symbolism

Subtraction is the opposite, or inverse, of addition. If you know how to add (2 + 3 = 5) and solve for an unknown (__ + 2 = 5 and __ + 3 = 5), then subtraction will follow easily. Rather than teach another complete set of facts, we will relate subtraction to these two previously mastered skills. Every subtraction problem can be rewritten and rephrased as an addition problem. We can read the example 5 – 2 = ? as "What number plus 2 is the same as 5?" This is identical to __ + 2 = 5. You can see this with the blocks in example 1 on the next page.

The answer to a subtraction problem is called the ***difference***. Instead of focusing on taking away or "minusing," focus instead on the difference between the two numbers being subtracted. That is why instead of taking away, we can add up.

Set this up exactly as an addition problem by pushing the blocks end to end. Since subtraction is the opposite of addition, invert the two bar and place it on top of the five bar. The dark blocks in the pictures represent the hollow side of each block. When the hollow side is showing, it means "***take away***" or "***minus***" or "owe." One parent commented that you are "in the hole" when the hollow side is showing.

In this lesson, work on rewriting a subtraction problem as an addition problem. Use the blocks to show how subtraction is the inverse of addition. Help the student see how each problem is solved by adding up. Since we are working in *Primer*, use the blocks in every problem. This is an introduction to a subject covered more fully in the succeeding two books.

Example 1

Solve 5 – 2 or $\begin{array}{r} 5 \\ \underline{-2} \end{array}$

Make a this a real problem by saying, "I have five dollars and I owe the paper boy two dollars. How many dollars do I have left?"

Step 1   Rephrase as an addition problem.

"Five minus two" is the same as
"What plus two is the same as five?"

Step 2   Rewrite as solving for an unknown.

_____ + 2 = 5

Step 3
Build as an addition problem.

Step 4
Invert and reposition.

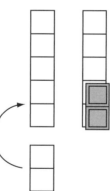

Step 5   Solve and write.

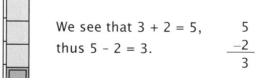

We see that 3 + 2 = 5, $\begin{array}{r} 5 \\ \underline{-2} \\ 3 \end{array}$
thus 5 – 2 = 3.

Example 2

Solve 7 - 3 or $\begin{array}{r} 7 \\ -3 \\ \hline \end{array}$

Step 1   Rephrase as an addition problem.

"Seven minus three" is the same as
"What plus three is the same as seven?"

Step 2   Rewrite as solving for an unknown.

_____ + 3 = 7

Step 3   **Step 4**
Build as an addition problem.   Invert and reposition.

   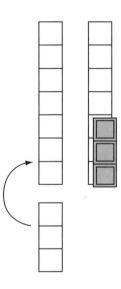

Step 5   Solve and write.

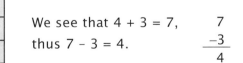

We see that 4 + 3 = 7, $\begin{array}{r} 7 \\ -3 \\ \hline 4 \end{array}$
thus 7 − 3 = 4.

Subtraction: −1

As with addition, we'll begin with the simpler facts, such as subtracting by one. See example 1.

Before teaching subtraction by one, you may want to review by counting backwards by one with the blocks arranged from one to nine as we did in lesson 13.

Example 1

Solve 4 − 1 or
$$\begin{array}{r} 4 \\ -1 \\ \hline \end{array}$$

Step 1   Rephrase as an addition problem.

"Four minus one" is the same as
"What plus one is the same as four?"

Step 2   Rewrite as solving for an unknown.

____ + 1 = 4

Step 3

Build as an addition problem.

Step 4

Invert and reposition.

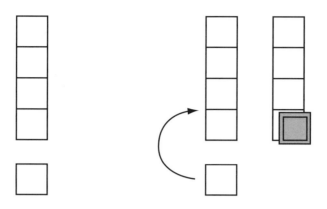

Step 5   Solve and write.

We see that 3 + 1 = 4,   4
thus 4 − 1 = 3.   −1
                            3

Student Solutions

Lesson Practice 1A
Count the blocks and circle the correct number.

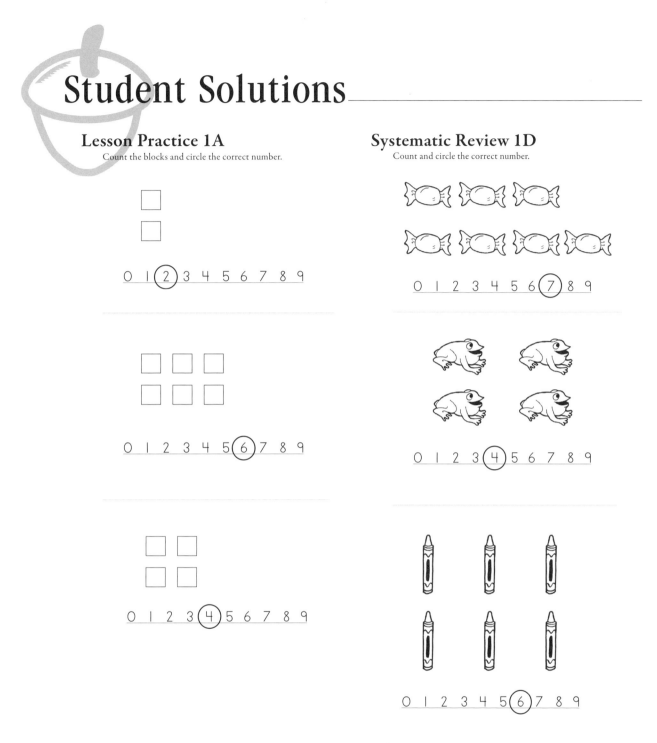

0 1 ②3 4 5 6 7 8 9

0 1 2 3 4 5 ⑥ 7 8 9

0 1 2 3 ④ 5 6 7 8 9

Systematic Review 1D
Count and circle the correct number.

0 1 2 3 4 5 6 ⑦ 8 9

0 1 2 3 ④ 5 6 7 8 9

0 1 2 3 4 5 ⑥ 7 8 9

Lesson Practice 2A
Cross out the wrong answer. Trace the correct answer.

Systematic Review 2D
Cross out the wrong answer. Trace the correct answer.

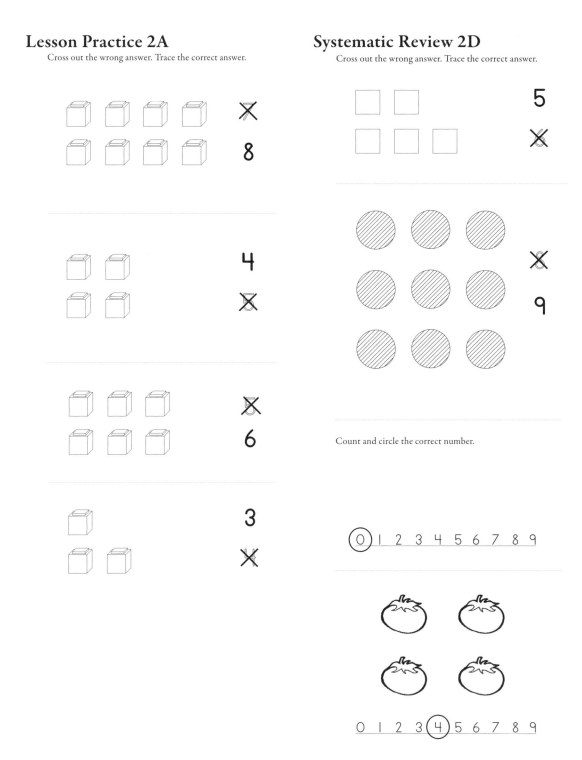

Count and circle the correct number.

Lesson Practice 3A

Shade or color the squares to show the correct number.
Trace the numerals.

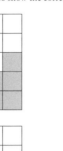

5

3

8

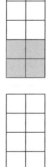

4

1

6

Systematic Review 3D

Shade or color the squares to show the correct number.
Trace the numerals.

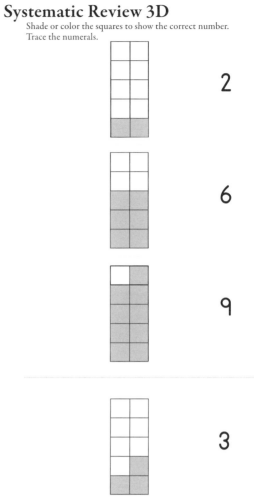

2

6

9

3

Cross out the wrong answer. Trace the correct answer.

5

Lesson Practice 4A

Count the rectangles that match the picture.
Circle and say the correct number.

How many rectangles are white? ☐

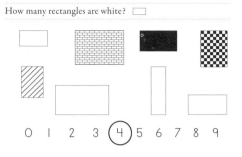

0 I 2 3 ④ 5 6 7 8 9

How many rectangles are black? ▬

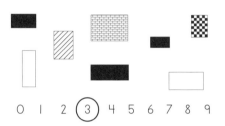

0 I 2 ③ 4 5 6 7 8 9

How many rectangles have stripes? ▨

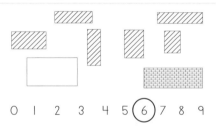

0 I 2 3 4 5 ⑥ 7 8 9

Systematic Review 4D

Count the rectangles that match the picture.
Circle and say the correct number.

How many rectangles have bricks?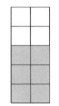

0 I 2 3 4 5 6 7 8 ⑨

Cross out the wrong answer. Trace the correct answer.

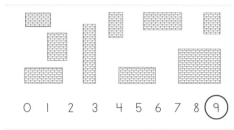

7

✗

Shade or color the squares to show the correct number.
Trace the numerals.

6

2

Lesson Practice 5A

Put the blocks on the squares. Count and say the correct number.
Write it if you can.

Systematic Review 5D

Put the blocks on the squares. Count and say the correct number.
Write it if you can.

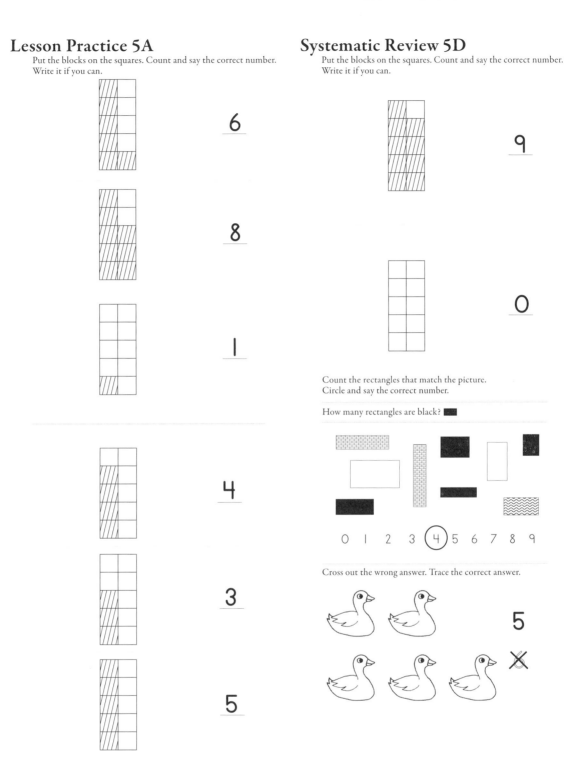

6

8

1

9

0

4

3

5

Count the rectangles that match the picture.
Circle and say the correct number.

How many rectangles are black? ▬

0 1 2 3 ④ 5 6 7 8 9

Cross out the wrong answer. Trace the correct answer.

5

Lesson Practice 6A

Count the circles that match the picture.
Circle and say the correct number.

How many circles are white? ○

0 1 2 3 4 5 6 (7) 8 9

How many circles are black? ●

0 1 (2) 3 4 5 6 7 8 9

How many circles have stripes? ⊘

0 1 2 3 4 (5) 6 7 8 9

Systematic Review 6D

Count the circles that match the picture.
Circle and say the correct number.

How many circles have stripes? ⊘

0 1 2 3 4 5 6 (7) 8 9

How many rectangles are black? ▬

0 1 2 3 (4) 5 6 7 8 9

Put the blocks on the squares. Count and say the correct number.
Write it if you can.

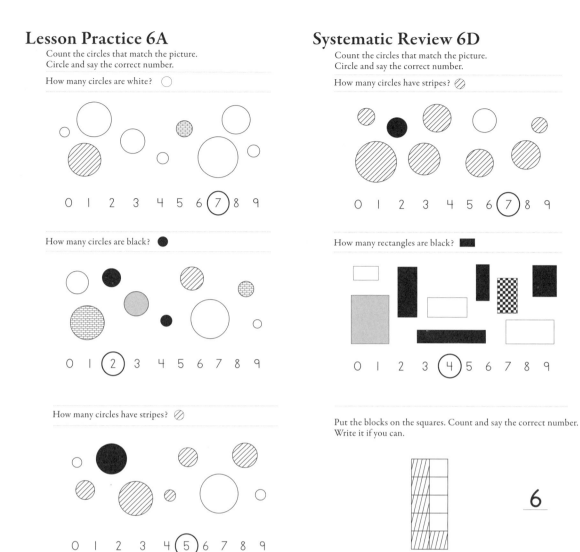

6

2

Lesson Practice 7A

Color the squares or write the number.

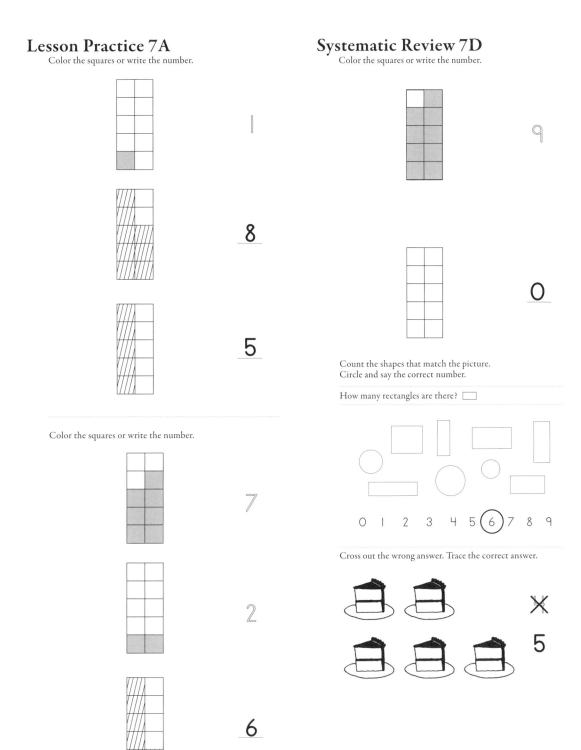

1

8

5

Color the squares or write the number.

7

2

6

Systematic Review 7D

Color the squares or write the number.

9

0

Count the shapes that match the picture.
Circle and say the correct number.

How many rectangles are there? ☐

0 1 2 3 4 5 ⑥ 7 8 9

Cross out the wrong answer. Trace the correct answer.

5

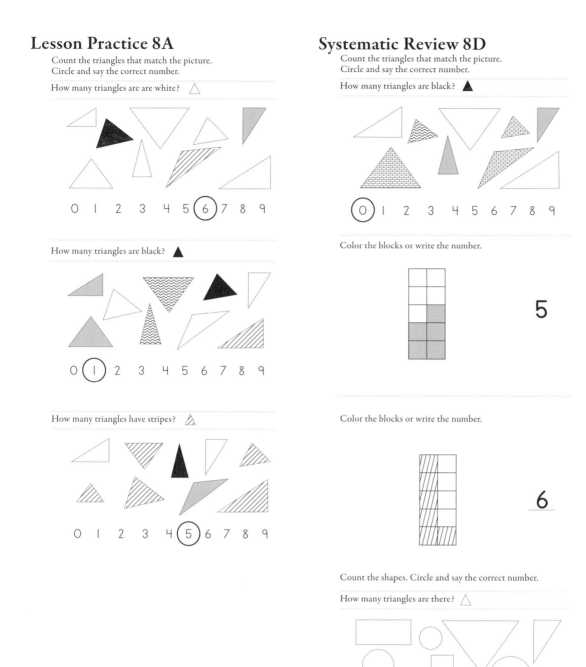

Lesson Practice 8A

Count the triangles that match the picture.
Circle and say the correct number.

How many triangles are are white? △

0 I 2 3 4 5 ⑥ 7 8 9

How many triangles are black? ▲

0 ① 2 3 4 5 6 7 8 9

How many triangles have stripes? ◸

0 I 2 3 4 ⑤ 6 7 8 9

Systematic Review 8D

Count the triangles that match the picture.
Circle and say the correct number.

How many triangles are black? ▲

⓪ I 2 3 4 5 6 7 8 9

Color the blocks or write the number.

5

Color the blocks or write the number.

<u>6</u>

Count the shapes. Circle and say the correct number.

How many triangles are there? △

0 I 2 ③ 4 5 6 7 8 9

Lesson Practice 9A

Color the correct number of blocks. Say the number.

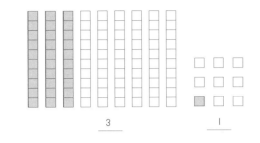

3   1

Count and write, and then say the number.

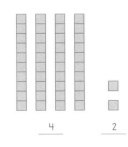

4   2

Build and say the numbers.

89

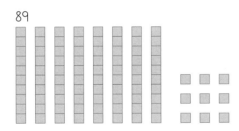

57

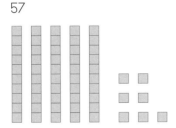

Systematic Review 9D

Color the correct number of blocks. Say the number.

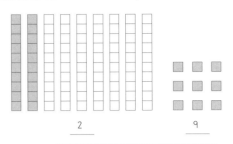

2   9

Build and say the numbers.

81

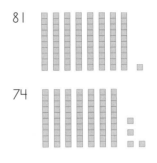

74

Count the shapes. Circle and say the correct number.

How many triangles are there? △

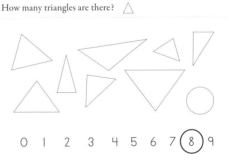

0 1 2 3 4 5 6 7 (8) 9

Lesson Practice 10A

Color the correct number of blocks. Say the number.

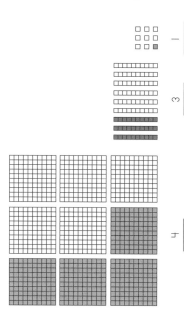

Systematic Review 10D

Color the correct number of blocks. Say the number.

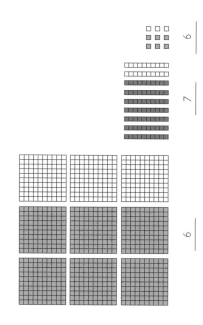

Count and write, and then say the number.

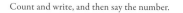

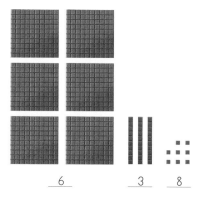

<u>6</u>   <u>3</u>   <u>8</u>

Count and write, and then say the number.

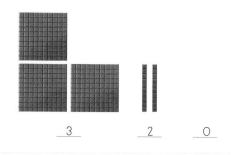

<u>3</u>   <u>2</u>   <u>0</u>

Count the rectangles. Circle and say the correct number. ▭

Ⓞ 1 2 3 4 5 6 7 8 9

Lesson Practice 11A

Put the unit bars on the pictures.
Count, write, and say the numbers.
Color the pictures to match the unit bars.

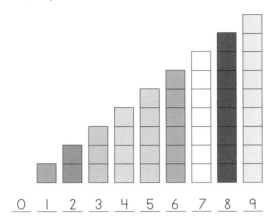

0 1 2 3 4 5 6 7 8 9

Use the blocks. Count, match, and color.
Write the number and say it.

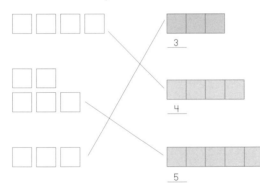

3

4

5

Systematic Review 11D

Use the blocks. Count, match, and color.
Write the number and say it.

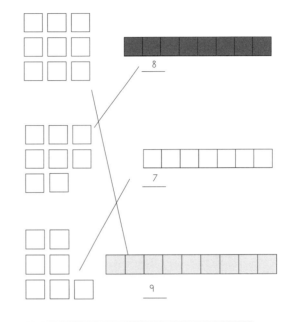

8

7

9

Count and write, and then say the number.

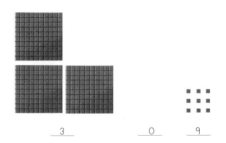

3   0   9

Lesson Practice 12A

Build each addition problem. Say it and write it.
Color the picture to match the unit bars.
You will need to turn your book sideways.

Systematic Review 12D

Build each addition problem.
Say it and write it. Color the
picture to match the unit bars.

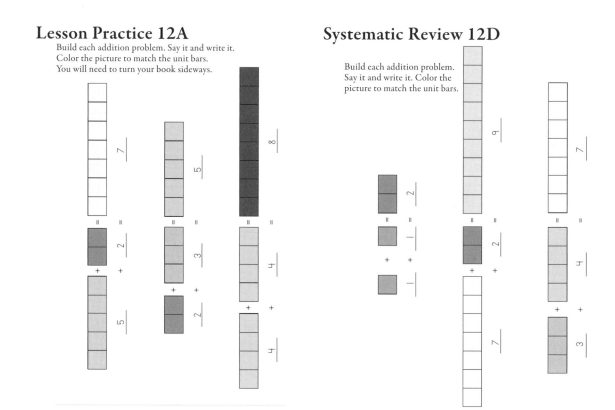

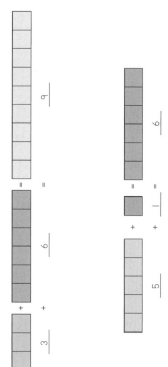

Count and write, and then say the number.

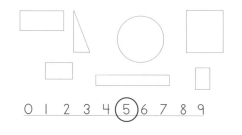

Count the rectangles. Circle and say the correct number. ▭

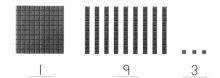

0 1 2 3 4 ⑤ 6 7 8 9

Lesson Practice 13A
Build, say, write, and color.

$\underline{1} + \underline{1} = \underline{2}$

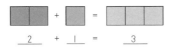

$\underline{2} + \underline{1} = \underline{3}$

$\underline{4} + \underline{1} = \underline{5}$

$\underline{5} + \underline{1} = \underline{6}$

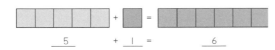

$\underline{3} + \underline{1} = \underline{4}$

Systematic Review 13D
Build, say, write, and color.

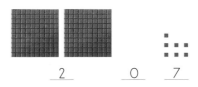

Count and write, and then say the number.

$\underline{2} \qquad \underline{0} \qquad \underline{7}$

Count the circles. Circle and say the correct number. ○

$0 \quad 1 \quad 2 \quad ③ \quad 4 \quad 5 \quad 6 \quad 7 \quad 8 \quad 9$

Lesson Practice 14A

Write the numbers that match the blocks. Some numbers are already written for you.

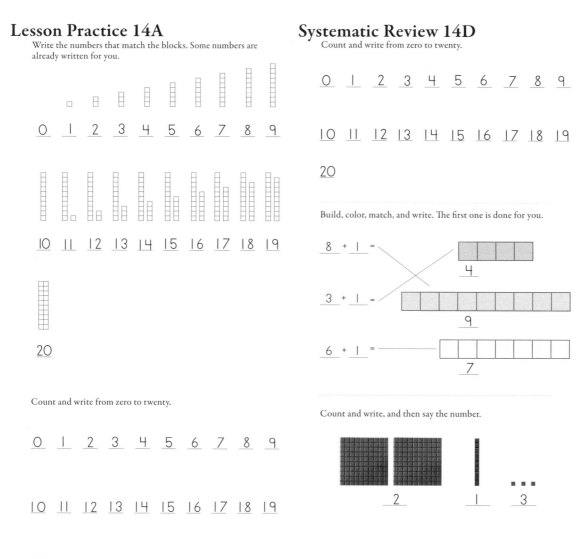

0 1 2 3 4 5 6 7 8 9

10 11 12 13 14 15 16 17 18 19

20

Count and write from zero to twenty.

0 1 2 3 4 5 6 7 8 9

10 11 12 13 14 15 16 17 18 19

20

Systematic Review 14D

Count and write from zero to twenty.

0 1 2 3 4 5 6 7 8 9

10 11 12 13 14 15 16 17 18 19

20

Build, color, match, and write. The first one is done for you.

$8 + 1 =$ 4

$3 + 1 =$ 9

$6 + 1 =$ 7

Count and write, and then say the number.

2   1   3

Lesson Practice 15A

Build, say, write, and color.

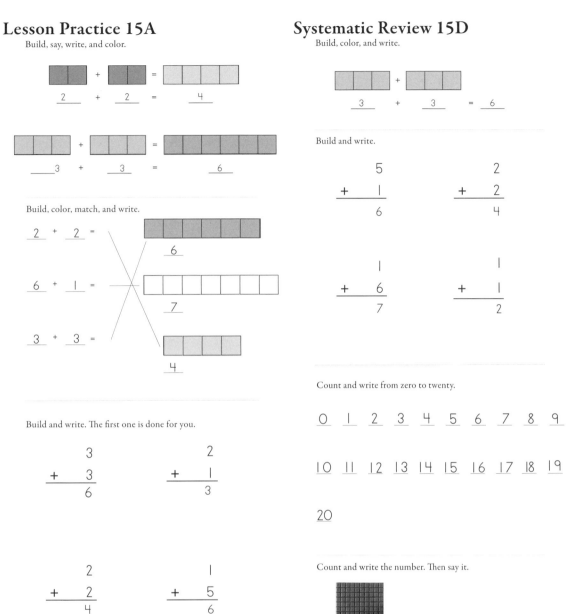

$$\underline{2} + \underline{2} = \underline{4}$$

$$\underline{\quad 3} + \underline{3} = \underline{6}$$

Build, color, match, and write.

$$\underline{2} + \underline{2} = \qquad \underline{6}$$

$$\underline{6} + \underline{1} = \qquad \underline{7}$$

$$\underline{3} + \underline{3} = \qquad \underline{4}$$

Build and write. The first one is done for you.

$$\begin{array}{r} 3 \\ + \ 3 \\ \hline 6 \end{array} \qquad \begin{array}{r} 2 \\ + \ 1 \\ \hline 3 \end{array}$$

$$\begin{array}{r} 2 \\ + \ 2 \\ \hline 4 \end{array} \qquad \begin{array}{r} 1 \\ + \ 5 \\ \hline 6 \end{array}$$

Systematic Review 15D

Build, color, and write.

$$\underline{3} + \underline{3} = \underline{6}$$

Build and write.

$$\begin{array}{r} 5 \\ + \ 1 \\ \hline 6 \end{array} \qquad \begin{array}{r} 2 \\ + \ 2 \\ \hline 4 \end{array}$$

$$\begin{array}{r} 1 \\ + \ 6 \\ \hline 7 \end{array} \qquad \begin{array}{r} 1 \\ + \ 1 \\ \hline 2 \end{array}$$

Count and write from zero to twenty.

$$\underline{0} \ \underline{1} \ \underline{2} \ \underline{3} \ \underline{4} \ \underline{5} \ \underline{6} \ \underline{7} \ \underline{8} \ \underline{9}$$

$$\underline{10} \ \underline{11} \ \underline{12} \ \underline{13} \ \underline{14} \ \underline{15} \ \underline{16} \ \underline{17} \ \underline{18} \ \underline{19}$$

$$\underline{20}$$

Count and write the number. Then say it.

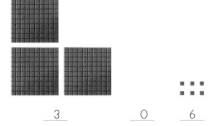

$$\underline{3} \qquad \underline{0} \qquad \underline{6}$$

Lesson Practice 16A

The diagrams are smaller than actual size. Have the student count to select the correct unit bars to complete the problems.

Build, say, write, and color.

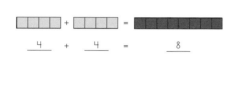

$$\underline{4} + \underline{4} = \underline{8}$$

$$\underline{5} + \underline{5} = \underline{10}$$

Build and write.

$$\begin{array}{r} 5 \\ + 5 \\ \hline 10 \end{array} \qquad \begin{array}{r} 4 \\ + 4 \\ \hline 8 \end{array}$$

$$\begin{array}{r} 1 \\ + 6 \\ \hline 7 \end{array} \qquad \begin{array}{r} 2 \\ + 2 \\ \hline 4 \end{array}$$

Count the squares that match the pictures.
Circle and say the correct number.

How many white squares are there? ▢

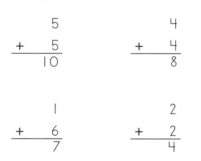

0 1 2 3 4 (5) 6 7 8 9

Systematic Review 16D

Build and write.

$$\begin{array}{r} 1 \\ + 8 \\ \hline 9 \end{array} \qquad \begin{array}{r} 4 \\ + 4 \\ \hline 8 \end{array}$$

$$\begin{array}{r} 2 \\ + 1 \\ \hline 3 \end{array} \qquad \begin{array}{r} 5 \\ + 5 \\ \hline 10 \end{array}$$

Count the squares that match the pictures.
Circle and say the correct number.

How many squares are gray? ▨

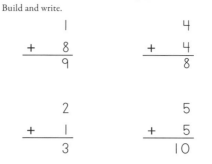

0 1 2 3 4 5 6 (7) 8 9

Count and write from zero to twenty.

$$\underline{0}\ \ \underline{1}\ \ \underline{2}\ \ \underline{3}\ \ \underline{4}\ \ \underline{5}\ \ \underline{6}\ \ \underline{7}\ \ \underline{8}\ \ \underline{9}$$

$$\underline{10}\ \ \underline{11}\ \ \underline{12}\ \ \underline{13}\ \ \underline{14}\ \ \underline{15}\ \ \underline{16}\ \ \underline{17}\ \ \underline{18}\ \ \underline{19}$$

$$\underline{20}$$

Build and say the numbers.

216

161

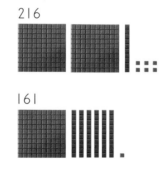

Lesson Practice 17A

How many squares are there?
Skip count by two and write the numbers on the lines.
Write the number of squares on the line outside the boxes.
The first one is done for you.

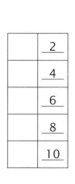

	2
	4
	6
	8
	10

10

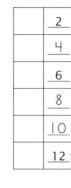

	2
	4
	6
	8
	10
	12

12

	2
	4
	6
	8
	10
	12
	14

14

	2
	4
	6
	8
	10
	12
	14

14

Skip count by two to find how many circles.

○ ○ ○ ○ ○
○ ○ ○ ○ ○   10 circles

Systematic Review 17D

How many squares are there? Skip count by two.
Write the numbers.

	2
	4
	6
	8
	10
	12
	14
	16
	18

18

	2
	4
	6
	8
	10
	12
	14
	16
	18
	20

20

Skip count by two to find how many chairs.

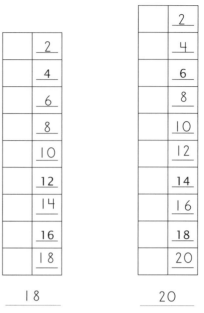

16 chairs

Build and write.

$$\begin{array}{r} 7 \\ +\ 1 \\ \hline 8 \end{array}$$

$$\begin{array}{r} 1 \\ +\ 4 \\ \hline 5 \end{array}$$

$$\begin{array}{r} 2 \\ +\ 2 \\ \hline 4 \end{array}$$

$$\begin{array}{r} 1 \\ +\ 1 \\ \hline 2 \end{array}$$

Lesson Practice 18A

Build and add the tens. Say the problem as you write it.

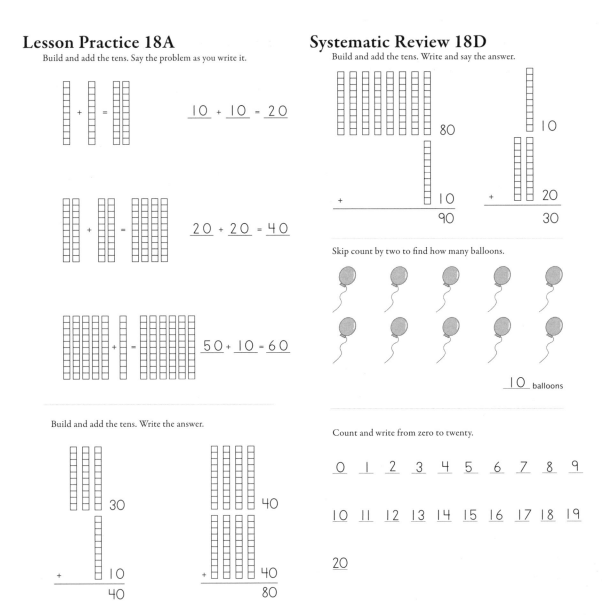

$10 + 10 = 20$

$20 + 20 = 40$

$50 + 10 = 60$

Build and add the tens. Write the answer.

30
+ 10
40

40
+ 40
80

Systematic Review 18D

Build and add the tens. Write and say the answer.

80
+ 10
90

10
+ 20
30

Skip count by two to find how many balloons.

10 balloons

Count and write from zero to twenty.

0 1 2 3 4 5 6 7 8 9

10 11 12 13 14 15 16 17 18 19

20

Lesson Practice 19A

Skip count by 10 and write the numbers on the lines.
Then write the numbers in the spaces under the squares.

							10
							20
							30
							40
							50

10, 20, 3 0, 4 0, 50

Skip count by 10 and write the numbers on the lines.
Then write the numbers in the spaces under the squares.

							10
							20
							30
							40
							50
							60
							70
							80
							90
							100

10, 20, 30, 40, 50, 60, 70, 80, 90, 100

Systematic Review 19D

Skip count by 10. Write the missing numbers on the lines.

10, 20, 30, 40, 50, 60, 70, 80, 90, 100

Skip count by 10 to find how many squares.

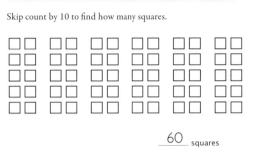

_60_ squares

Build and write.

$10 + 10 = \underline{20}$ $4 + 1 = \underline{5}$

$$\begin{array}{r} 1 \\ +\ 7 \\ \hline 8 \end{array}$$ $$\begin{array}{r} 2 \\ +\ 2 \\ \hline 4 \end{array}$$

$$\begin{array}{r} 2\ 0 \\ +\ 1\ 0 \\ \hline 3\ 0 \end{array}$$ $$\begin{array}{r} 4\ 0 \\ +\ 4\ 0 \\ \hline 8\ 0 \end{array}$$

Read the word problem to the student if necessary. Use the pictures to help solve it.

Pam has three pieces of cake. Pat has one piece of cake.
How much cake do Pam and Pat have in all?

 = _4_ pieces

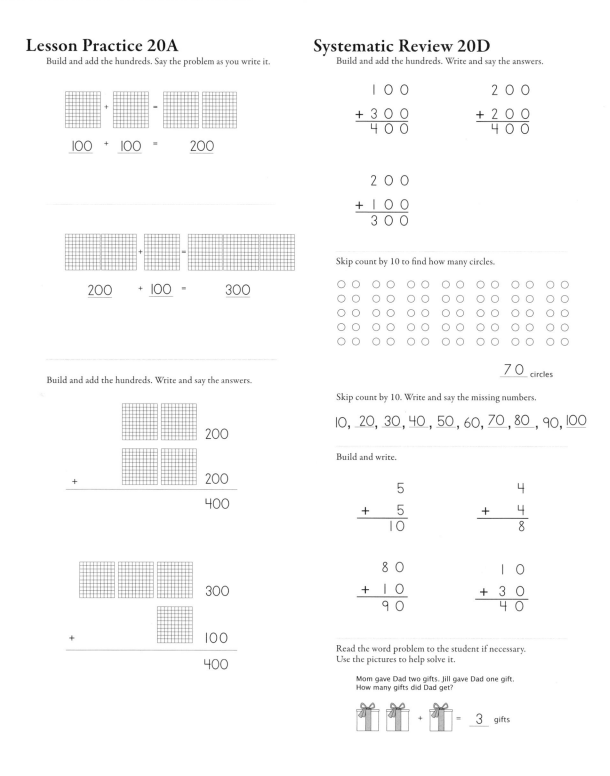

Lesson Practice 20A

Build and add the hundreds. Say the problem as you write it.

100 + 100 = 200

200 + 100 = 300

Build and add the hundreds. Write and say the answers.

200
+ 200
400

300
+ 100
400

Systematic Review 20D

Build and add the hundreds. Write and say the answers.

$$\begin{array}{r} 1\ 0\ 0 \\ +\ 3\ 0\ 0 \\ \hline 4\ 0\ 0 \end{array}$$

$$\begin{array}{r} 2\ 0\ 0 \\ +\ 2\ 0\ 0 \\ \hline 4\ 0\ 0 \end{array}$$

$$\begin{array}{r} 2\ 0\ 0 \\ +\ 1\ 0\ 0 \\ \hline 3\ 0\ 0 \end{array}$$

Skip count by 10 to find how many circles.

70 circles

Skip count by 10. Write and say the missing numbers.

10, 20, 30, 40, 50, 60, 70, 80, 90, 100

Build and write.

$$\begin{array}{r} 5 \\ +\ 5 \\ \hline 1\ 0 \end{array}$$

$$\begin{array}{r} 4 \\ +\ 4 \\ \hline 8 \end{array}$$

$$\begin{array}{r} 8\ 0 \\ +\ 1\ 0 \\ \hline 9\ 0 \end{array}$$

$$\begin{array}{r} 1\ 0 \\ +\ 3\ 0 \\ \hline 4\ 0 \end{array}$$

Read the word problem to the student if necessary.
Use the pictures to help solve it.

Mom gave Dad two gifts. Jill gave Dad one gift.
How many gifts did Dad get?

= 3 gifts

Lesson Practice 21A

Use the blocks to solve for the unknown. Write the answer in the blank and say it.

_2_ + 2 = 4

_3_ + 3 = 6

_6_ + 1 = 7

_5_ + 5 = 10

_3_ + 1 = 4

_1_ + 4 = 5

Read the word problem. Use the blocks to help you solve for the unknown.

Nick wants three toy airplanes. He has two airplanes. How many more airplanes does Nick want?

Systematic Review 21D

Use the blocks to solve for the unknown. Write the answer in the blank and say it.

_4_ + 2 = 6

_5_ + 5 = 10

Build and write.

3 + 1 = _4_

10 + 60 = _70_

Build and write.

```
   1          2
+  8       +  2
   9          4

  3 0       1 0 0
+ 1 0      +1 0 0
  4 0       2 0 0
```

Skip count by two to find how many triangles.

_18_ triangles

Read the word problem. Use the blocks to help you solve for the unknown.

Jed and Jack have two puppies. One puppy belongs to Jed. How many puppies belong to Jack?

_1_ +

Lesson Practice 22A

Skip count by five and write the numbers on the lines.
The first one is done for you.

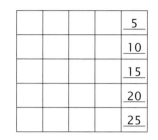

				5
				10
				15
				20
				25

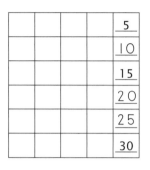

				5
				10
				15
				20
				25
				30

Skip count by five and write the numbers on the lines.
Then write the numbers in the spaces under the squares.

				5
				10
				15
				20
				25
				30
				35
				40

5, 10, 15, 20, 25, 30, 35, 40

Systematic Review 22D

Skip count by five and write the numbers on the lines.
Then write the numbers in the spaces under the squares.

				5
				10
				15
				20
				25
				30
				35
				40
				45
				50

5, 10, 15, 20, 25, 30, 35, 40, 45, 50

Skip count by five to find how many balloons.

_20_ balloons

Use the blocks to solve for the unknown. Write the answer
in the blank and say it.

_7_ + 2 = 9

_1_ + 4 = 5

Lesson Practice 23A

Change the tally marks to a number. The first one is done for you.

꘡꘡꘡ = __5__

꘡꘡꘡ ꘡꘡꘡ = __10__

꘡꘡꘡ ꘡ = __6__

꘡꘡꘡ ꘡꘡꘡ ꘡ = __11__

꘡꘡꘡ ꘡꘡꘡ ꘡꘡꘡ ꘡꘡꘡ = __20__

꘡꘡꘡ ꘡꘡꘡ ꘡꘡꘡ ꘡꘡꘡ ꘡꘡꘡ ꘡ = __26__

Use tally marks to write the number. The first one is done for you.

17 = __꘡꘡꘡ ꘡꘡꘡ ꘡꘡꘡ ꘡꘡__

10 = __꘡꘡꘡ ꘡꘡꘡__

6 = __꘡꘡꘡ ꘡__

15 = __꘡꘡꘡ ꘡꘡꘡ ꘡꘡꘡__

Systematic Review 23D

Change the tally marks to a number.

꘡꘡꘡ ꘡꘡꘡ ꘡꘡꘡ ꘡ = __16__

Use tally marks to write the number.

10 = __꘡꘡꘡ ꘡꘡꘡__

Skip count by five. Write the missing numbers on the lines.

5, __10__, 15, __20__, __25__, 30, __35__, __40__, __45__, 50

Use the blocks to solve for the unknown. Write the answer in the blank and say it.

__1__ + 7 = 8

__2__ + 3 = 5

Build and write.

$$\begin{array}{r} 6 \\ +\ 1 \\ \hline 7 \end{array}$$
$$\begin{array}{r} 4 \\ +\ 4 \\ \hline 8 \end{array}$$

$$\begin{array}{r} 50 \\ +\ 10 \\ \hline 60 \end{array}$$
$$\begin{array}{r} 100 \\ +200 \\ \hline 300 \end{array}$$

Lesson Practice 24A

The diagrams are smaller than actual size. Have the student count to select the correct unit bars to complete the problems.

Build, say, write, and colour.

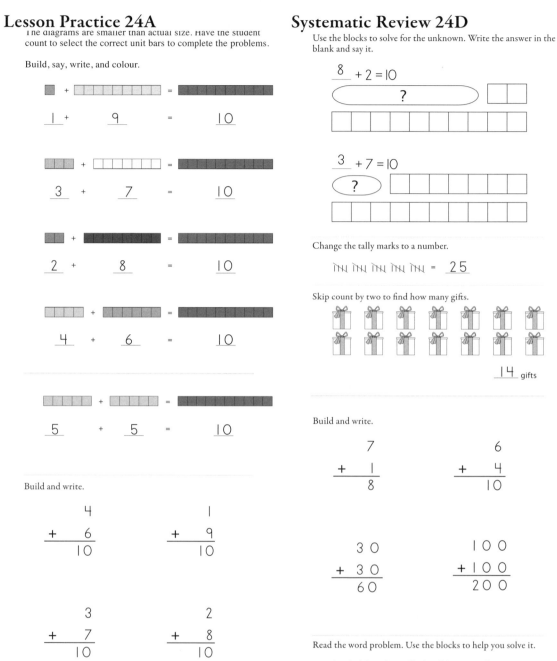

$\underline{1}$ + $\underline{9}$ = $\underline{10}$

$\underline{3}$ + $\underline{7}$ = $\underline{10}$

$\underline{2}$ + $\underline{8}$ = $\underline{10}$

$\underline{4}$ + $\underline{6}$ = $\underline{10}$

$\underline{5}$ + $\underline{5}$ = $\underline{10}$

Build and write.

$$\begin{array}{r} 4 \\ + \ 6 \\ \hline 10 \end{array} \qquad \begin{array}{r} 1 \\ + \ 9 \\ \hline 10 \end{array}$$

$$\begin{array}{r} 3 \\ + \ 7 \\ \hline 10 \end{array} \qquad \begin{array}{r} 2 \\ + \ 8 \\ \hline 10 \end{array}$$

Systematic Review 24D

Use the blocks to solve for the unknown. Write the answer in the blank and say it.

$\underline{8}$ + 2 = 10

?

$\underline{3}$ + 7 = 10

?

Change the tally marks to a number.

𝍷𝍷𝍷𝍷 𝍷𝍷𝍷𝍷 𝍷𝍷𝍷𝍷 𝍷𝍷𝍷𝍷 𝍷𝍷𝍷𝍷 = $\underline{25}$

Skip count by two to find how many gifts.

$\underline{14}$ gifts

Build and write.

$$\begin{array}{r} 7 \\ + \ 1 \\ \hline 8 \end{array} \qquad \begin{array}{r} 6 \\ + \ 4 \\ \hline 10 \end{array}$$

$$\begin{array}{r} 30 \\ + \ 30 \\ \hline 60 \end{array} \qquad \begin{array}{r} 100 \\ +100 \\ \hline 200 \end{array}$$

Read the word problem. Use the blocks to help you solve it.

Pam had three dimes. She found three more dimes. How many dimes does Pam have now?

3 + 3 = $\underline{6}$ dimes

Lesson Practice 25A

Find the area of each rectangle by skip counting.
Write the numbers on the lines if you need to.
Write your answer in the last box.

```
___
___
___
___
___
3 0
```

```
___
___
___
___
1 0
```

Find the area of each rectangle by skip counting.
Write the numbers on the lines if you need to.
Write your answer in the last box.

```
___
___
___
___
___
___
___
8 0
```

Systematic Review 25D

Find the area of each rectangle by skip counting.
Write the numbers on the lines if you need to.
Write your answer in the last box.

```
___
___
8
```

```
___
6
```

```
___
___
___
1 0
```

Change the tally marks to a number.

卌 卌 卌 = 15

Build and write.

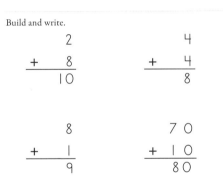

$$\begin{array}{r} 2 \\ + 8 \\ \hline 1 0 \end{array}$$

$$\begin{array}{r} 4 \\ + 4 \\ \hline 8 \end{array}$$

$$\begin{array}{r} 8 \\ + 1 \\ \hline 9 \end{array}$$

$$\begin{array}{r} 7 0 \\ + 1 0 \\ \hline 8 0 \end{array}$$

Read the word problem. Use the blocks to help you solve it.

Four birds sang to me. One more bird came to sing. How many birds are singing now?

4 + 1 = 5 birds

Lesson Practice 26A

Count the minutes. There is a removable clock template near the end of this book.

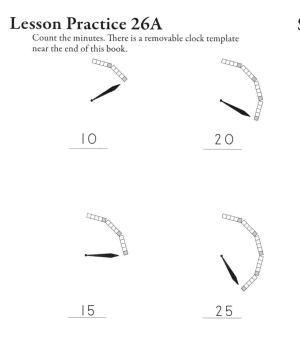

10

20

15

25

Write how many minutes are shown by each clock.

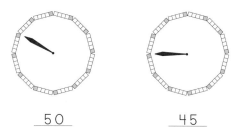

50

45

Systematic Review 26D

Write how many minutes are shown by each clock.

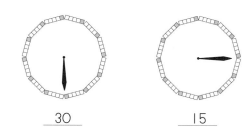

30

15

Find the area of each rectangle by skip counting. Write your answer in the last box.

4

12

8

Change the tally marks to a number.

𝍷𝍷𝍷 𝍷𝍷𝍷 𝍷𝍷𝍷 𝍷 = 16

Build and write.

$$\begin{array}{r} 3 \\ + 3 \\ \hline 6 \end{array}$$

$$\begin{array}{r} 4 \\ + 6 \\ \hline 10 \end{array}$$

$$\begin{array}{r} 8 \\ + 1 \\ \hline 9 \end{array}$$

$$\begin{array}{r} 20 \\ + 10 \\ \hline 30 \end{array}$$

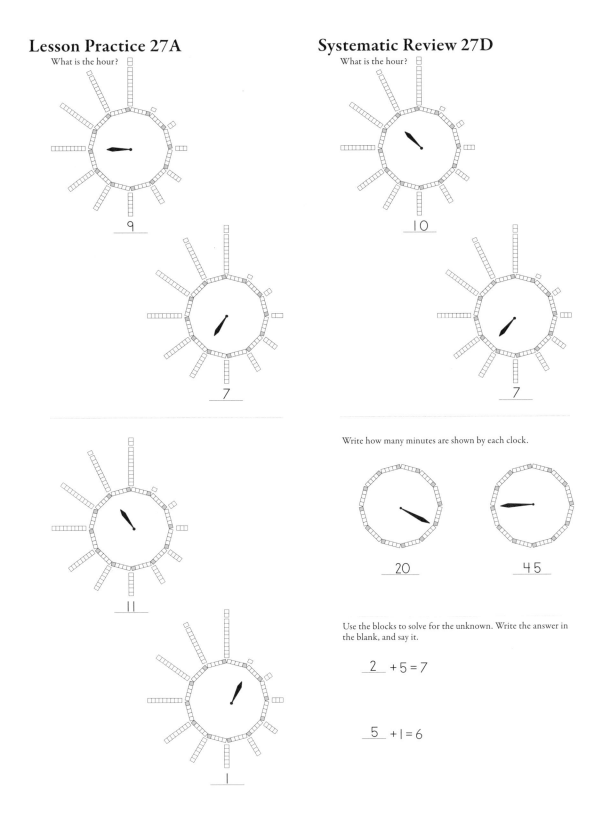

Lesson Practice 27A

What is the hour?

9

7

11

1

Systematic Review 27D

What is the hour?

10

7

Write how many minutes are shown by each clock.

20

45

Use the blocks to solve for the unknown. Write the answer in the blank, and say it.

_2_ + 5 = 7

_5_ + 1 = 6

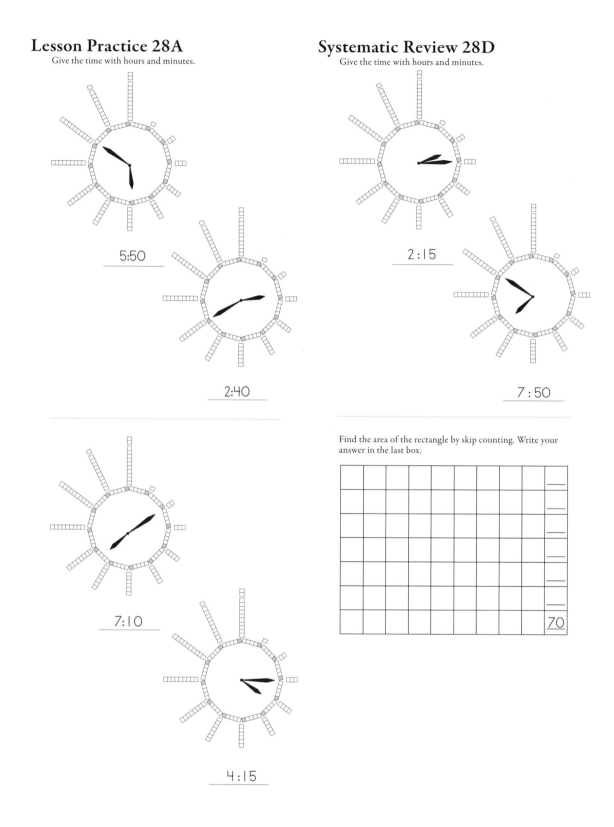

Lesson Practice 28A

Give the time with hours and minutes.

5:50

2:40

7:10

4:15

Systematic Review 28D

Give the time with hours and minutes.

2:15

7:50

Find the area of the rectangle by skip counting. Write your answer in the last box.

70

Lesson Practice 29A

Build each problem, and write the numbers in the circles.
The first one is done for you.

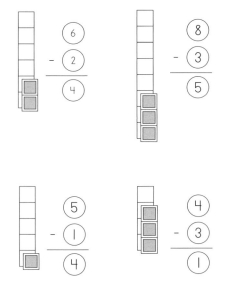

$$\begin{array}{r} 6 \\ - 2 \\ \hline 4 \end{array}$$

$$\begin{array}{r} 8 \\ - 3 \\ \hline 5 \end{array}$$

$$\begin{array}{r} 5 \\ - 1 \\ \hline 4 \end{array}$$

$$\begin{array}{r} 4 \\ - 3 \\ \hline 1 \end{array}$$

Use the blocks to build each problem. Say and write the answer.

$$\begin{array}{r} 4 \\ - 1 \\ \hline 3 \end{array}$$

$$\begin{array}{r} 5 \\ - 3 \\ \hline 2 \end{array}$$

Systematic Review 29D

Use the blocks to build each problem. Say and write the answer.

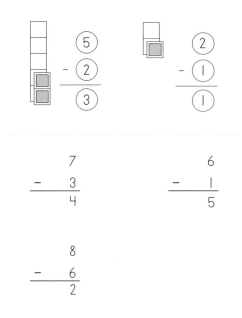

$$\begin{array}{r} 5 \\ - 2 \\ \hline 3 \end{array}$$

$$\begin{array}{r} 2 \\ - 1 \\ \hline 1 \end{array}$$

$$\begin{array}{r} 7 \\ - 3 \\ \hline 4 \end{array}$$

$$\begin{array}{r} 6 \\ - 1 \\ \hline 5 \end{array}$$

$$\begin{array}{r} 8 \\ - 6 \\ \hline 2 \end{array}$$

Give the time with hours and minutes.

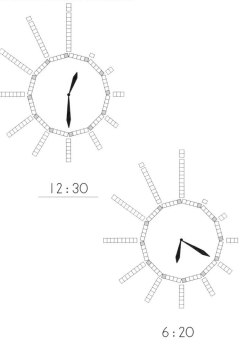

12 : 30

6 : 20

Lesson Practice 30A

Build each problem, and write the numbers in the circles.

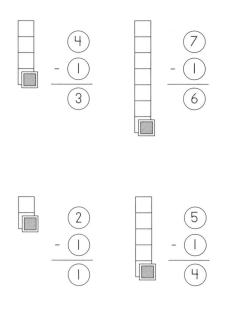

$$
\begin{array}{r}
4 \\
- 1 \\
\hline 3
\end{array}
\qquad
\begin{array}{r}
7 \\
- 1 \\
\hline 6
\end{array}
$$

$$
\begin{array}{r}
2 \\
- 1 \\
\hline 1
\end{array}
\qquad
\begin{array}{r}
5 \\
- 1 \\
\hline 4
\end{array}
$$

Use the blocks to build each problem. Say and write the answer.

$$
\begin{array}{r}
9 \\
- 1 \\
\hline 8
\end{array}
\qquad
\begin{array}{r}
6 \\
- 1 \\
\hline 5
\end{array}
$$

Systematic Review 30D

Use the blocks to build each problem. Say and write the answer.

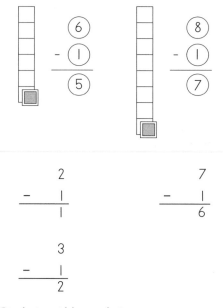

$$
\begin{array}{r}
6 \\
- 1 \\
\hline 5
\end{array}
\qquad
\begin{array}{r}
8 \\
- 1 \\
\hline 7
\end{array}
$$

$$
\begin{array}{r}
2 \\
- 1 \\
\hline 1
\end{array}
\qquad
\begin{array}{r}
7 \\
- 1 \\
\hline 6
\end{array}
$$

$$
\begin{array}{r}
3 \\
- 1 \\
\hline 2
\end{array}
$$

Give the time with hours and minutes.

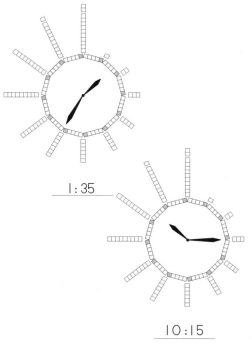

1:35

10:15

Master Index for General Math

This index lists the levels at which main topics are presented in the instruction manuals for *Primer* through *Zeta*. For more detail, see the description of each level at www.MathUSee.com. (Many of these topics are also reviewed in subsequent student books.)